Smithian (Early Triassic) ammonoid faunas from northwestern Guangxi (South China): taxonomy and biochronology

by

Arnaud Brayard and Hugo Bucher

Acknowledgement

Financial support for the publication of this number of Fossils and Strata was provided by the Swiss National Science Foundation.

Contents

Smithian (Early Triassic) ammonoid faunas from northwestern Guangxi (South China): taxonomy and biochronology

ARNAUD BRAYARD AND HUGO BUCHER

Brayard, A. & Bucher, H. 2008: Smithian (Early Triassic) ammonoid faunas from northwestern Guangxi (South China): taxonomy and biochronology. *Fossils and Strata*, No. 55, pp. 1–179. ISSN 0024-1164.

Intensive sampling of the Luolou Formation in northwestern Guangxi (South China) has led to the recognition of several new ammonoid faunas of Smithian age and to the construction of a new biostratigraphical zonation for the Smithian in the palaeoequatorial region. These faunas significantly enlarge the scope of the Smithian Stage, and the new zonal scheme facilitates correlation with other mid- and high-palaeolatitude faunal successions (i.e. British Columbia and Siberia). In ascending order, the new biostratigraphical sequence comprises: the *Clypites* sp. indet. beds of uppermost Dienerian age, the *Kashmirites kapila* beds, the *Flemingites rursiradiatus* beds, the *Owenites koeneni* beds, and the *Anasibirites multiformis* beds of Smithian age. Thus, the Smithian of this palaeoequatorial region now includes a newly introduced lowermost subdivision (i.e. the *Kashmirites kapila* beds) that may approximately be correlative of the *Hedenstroemia hedenstroemi* Zone of British Columbia and Siberia. Likewise, the newly introduced uppermost subdivision is equivalent to the *Anawasatchites tardus* Zone of British Columbia and Siberia.

Fourteen new genera (*Guangxiceltites, Weitschaticeras, Hebeisenites, Jinyaceras, Xiaoqiaoceras, Nanningites, Wailiceras, Leyeceras, Urdyceras, Galfettites, Guangxiceras, Larenites, Guodunites* and *Procurvoceratites*) and 36 new species (*Kashmirites guangxiense, Hanielites gracilus, H. angulus, Xenoceltites variocostatus, X. pauciradiatus, Guangxiceltites admirabilis, Weitschaticeras concavum, Hebeisenites varians, H. evolutus, H. compressus, Jinyaceras bellum, Juvenites procurvus, Paranorites jenksi, Xiaoqiaoceras involutus, Wailiceras aemulus, Leyeceras rothi, Urdyceras insolitus, Rohillites bruehwileri, R. sobolevi, Galfettites simplicitatis, Pseudoflemingites goudemandi, Guangxiceras inflata, Anaflemingites hochulii, Arctoceras strigatus, Hemiprionites klugi, Inyoites krystyni, 'Paranannites' ovum, P. subangulosus, P. dubius, Hedenstroemia augusta, Cordillerites antrum, Pseudaspenites evolutus, Guodunites monneti, Procurvoceratites pygmaeus, P. ampliatus* and *P. subtabulatus*) are described. □ Ammonoids, Early Triassic, Luolou Fm, northwestern Guangxi, Smithian, South China.

Arnaud Brayard [arnaud.brayard@univ-lyon1.fr], UMR 5125 PEPS CNRS, France; Université Lyon 1, Campus de la Doua, Bât. Géode, 69622 Villeurbanne Cedex, France; Present address: [arnaud.brayard@lmtg.obs-mip.fr], LMTG, UMR 5563 CNRS – Université Toulouse – IRD, Observatoire Midi-Pyrénées, 14 Avenue Edouard Belin, F-31400 Toulouse, France; Hugo Bucher [hugo.fr.bucher@pim.uzh.ch], Paläontologisches Institut und Museum der Universität Zürich, Karl-Schmid Strasse 4, CH-8006 Zürich, Switzerland.

Introduction

The marine Permo–Triassic boundary record is well preserved in South China and has attracted the attention of many scientists ever since the Meishan section was chosen as the global stratotype section and point (GSSP) (Yin *et al.* 1996, 2001). Moreover, Lower Triassic marine sedimentary formations are also widespread in the Guangxi and Guizhou provinces of South China. The pioneer contributions of Chao (1950, 1959) first documented the occurrence of rich Early Triassic ammonoid faunas in northwestern Guangxi. Chao (1950, 1959) directly integrated his data within the Flemingitan and Owenitan subdivisions of the biostratigraphical scheme of Spath (1934). Since Chao's studies, few papers have been published on the Early Triassic ammonoids of South China, and none has reassessed the Smithian ammonoid succession from northwestern Guangxi.

In order to better understand the dynamics of the biotic recovery following the Permo–Triassic mass extinction, Early Triassic palaeobiogeographical and diversity studies are now receiving more attention (e.g. Brayard *et al.* 2004, 2006, 2007b; Galfetti *et al.* 2007b,c). From this perspective, the South Chinese record is of prime importance. Indeed, marine Early

Triassic palaeoequatorial sections are few (see Brayard *et al.* 2006), thus further emphasizing the importance of the abundant ammonoid data from South China. Moreover, the widely accepted palaeoequatorial position of the South China Block during the Early Triassic makes it a key biogeographical reference, since the majority of recent works dealing with Early Triassic ammonoids are from either from mid- or high-palaeolatitudinal settings (e.g. Dagys & Ermakova 1990; Tozer 1994).

Our investigations in northwestern Guangxi have revealed several new Early Triassic faunas. In this paper we mainly focus on the taxonomy and biostratigraphy of the Smithian ammonoid faunas, as well as their implication for low palaeolatitude diversity and biochronology. Global correlations across latitudes and between both sides of Panthalassa are also discussed, as are new palaeobiogeographical and phylogenetic implications.

Geological framework

General context

Southwestern Asia is a complex collage of orogenic belts and successive accretions of Gondawana-derived continental blocks (Boulin 1991). China is tectonically complex and composed of several blocks that travelled across the Tethys during the Permian and Triassic (Enkin *et al.* 1992). Southwestern China is located at the junction of the Chinese, South-East Asian, and Indian plates, as well as the Tibetan block (Fan 1978), but the kinematic and temporal frame of their amalgamation is not well understood (Gilder *et al.* 1995). The South China Block and North China Block were located in the low latitudes of the eastern Tethys during the Early Triassic. Palaeomagnetic data indicate that the South China Block was situated near the Equator (Gilder *et al.* 1995).

Triassic deposits in South China

Marine Triassic sediments, represented by carbonate platforms and deep-water facies, are widespread in the South China Block and other neighbouring blocks. They are especially well developed and exposed in Tibet, Qinghai, Yunnan, Guizhou, Guangxi and Sichuan (Hsü 1940, 1943; Chao 1959; Wang *et al.* 1981). Generally, in the South China Block, the Lower Triassic and Anisian are represented by marine deposits, while the Upper Triassic is composed of continental sediments (Tong & Yin 2002).

Marine deposits in northwestern Guangxi belong to the Youjiang Sedimentary Province, which was part of the South China Block during the Early Triassic. Sedimentary deposits in this province, which is also known as the Nanpanjiang Basin, consist of clastic and carbonate rocks deposited in a deep-water basin with a few, smaller, isolated carbonate platforms distributed within Guangxi and Guizhou provinces (Lehrmann *et al.* 2003). The basin is bordered to the north and west by the large Yangtze carbonate platform, and Lower Triassic deposits are generally distributed along the northern and western edges of this platform.

The Luolou Formation

The Luolou Formation, which represents an important part of the Early Triassic outer platform facies in southern Guizhou and northwestern Guangxi, was first well described by Chao (1950, 1959). Its lower part mainly consists of dark grey, thin limestone beds alternating with dark shales, whereas its upper part is composed of massive, grey, nodular limestone. Overlying the Luolou Formation is the +1,000-m-thick Baifeng Formation of Anisian age, which consists mainly of clastic turbidites. This influx of terrigeneous material suggests a drowning of the basin at that time, and thus, a concomitant modification in directions or rates of convergence between the South and North China blocks (Gilder *et al.* 1995).

Sections from northwestern Guangxi

All the studied sections are located in northwestern Guangxi (Fig. 1). The Luolou Formation is located on the periphery of the Carboniferous–Permian karstic massifs, and often in fault contact with Middle Triassic siliciclastics. Classic sections such as Tsoteng, sampled by Chao (1950, 1959), were resampled in order to obtain a precise, detailed record, since bed-by-bed sampling was not systematically utilized for the original collections. The type section near Luolou is now poorly exposed and therefore, was not resampled. With the exception of Tsoteng, all sampled sections (Figs 2–12) yielding numerous fossiliferous layers were first correlated by means of lithological or marker beds (e.g. the 3 m thick limestone band containing the *Flemingites rursiradiatus* beds), which then facilitated the construction of a composite range chart for ammonoid faunas (Fig. 13).

The Smithian lithological succession is very similar within the Jinya and Leye areas and can be easily summarized (see also Fig. 14). Dark shales alternating with thin, laminated micritic limestone beds (Figs 3–5, 14 for details; Galfetti *et al.* 2007a) characterize the lower portion of the Smithian. These recessive rocks are usually only partly exposed and ammonoids are

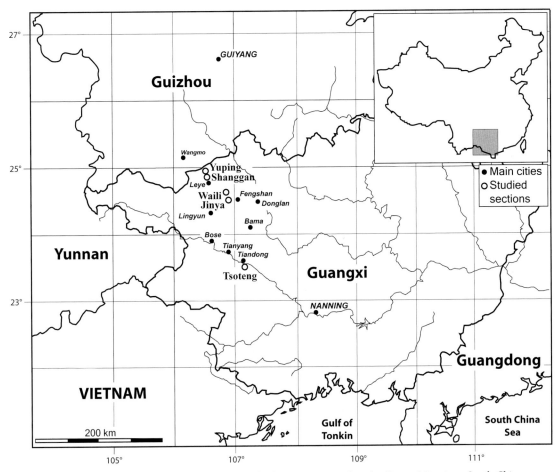

Fig. 1. Location map of sampled sections (Tsoteng, Jinya, Waili, Shanggan, Yuping) in the Guangxi Province, South China.

relatively rare. At best, the position of the Dienerian–Smithian boundary is bracketed within a 3-m-thick interval as shown in the Waili section (Figs 7, 14). These lowermost beds are overlain by a conspicuous, ca. 3 m thick, grey, thin-bedded limestone unit. This unit contains the *Flemingites rursiradiatus* beds, and it represents an important lithological marker found in all sections. This marker is overlain by the *Owenites koeneni* beds, which consist of dark laminated micritic limestone beds intercalated within dark shales. Overlying these beds are reddish weathering, dark carbonate siltstones, which contain the *Anasibirites multiformis* beds. The uppermost few metres consist of black shales containing small sized, early diagenetic limestone nodules yielding a few plant remains and typical *Xenoceltites* of latest Smithian age. Finally, the Smithian/Spathian boundary is characterized by an abrupt change to carbonate deposition (Fig. 14A). Contrasting with those of the Leye and Jinya areas, rocks of Smithian age in the Tsoteng area are almost exclusively composed of limestone.

Biostratigraphy

General subdivisions

Although several different schemes have been proposed for stage subdivision of the Lower Triassic, near unanimous agreement on any particular scheme remains somewhat elusive. The Lower Triassic is commonly divided into two, three or four substages as follows: two substages (Induan and Olenekian; the Olenekian encompassing the Smithian and the Spathian); three substages (Griesbachian, Nammalian and Spathian); or four substages (Griesbachian, Dienerian, Smithian and Spathian). In this paper we use the four substage subdivision proposed by Tozer (1967), whose boundaries are well defined in terms of ammonoids.

Chinese subdivisions

A Chinese stratigraphy commission initially proposed a twofold subdivision for the Lower Triassic, based

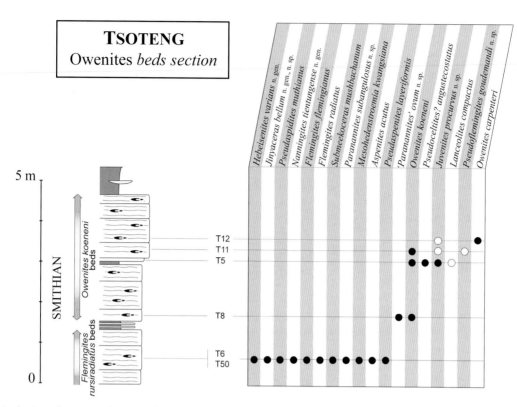

Fig. 2. Distribution of ammonoid taxa in the Tsoteng section. Open dots indicate occurrences based only on fragmentary or poorly preserved material.

on lithological differences (Tong *et al.* 2001; Tong & Yin 2002). Recently, the Chaohu section (Anhui Province, eastern China) was proposed as a GSSP candidate for the Induan/Olenekian boundary (IOB), with a position based on (i) the FAD of the conodont *Neospathodus waageni* Sweet and (ii) below the FAD of the Smithian ammonoids *Flemingites* and *Euflemingites* (Tong *et al.* 2004). In the Tethyan areas, the association of the ammonoids *Flemingites* and *Euflemingites* is usually recognized as the lowermost fauna of Smithian stage. Yet, these two flemingitids do not actually represent the oldest Smithian fauna (e.g. Krystyn *et al.* 2007). Therefore, the lowermost ammonoid faunas of the Smithian were not included in the original Chinese proposal. Thus, the lowest zone of the Smithian in the mid- and high-palaeolatitudes, i.e. the *Hedenstroemia hedenstroemi* Zone of British Columbia and Siberia (Tozer 1994; Ermakova 2002; Fig. 15) could not be easily correlated with localities in the equatorial palaeolatitudes (Shevyrev 2001, 2006).

Focusing only on the Smithian, the previous South Chinese subdivisions (e.g. Tong & Yin 2002) require some revision, since they were essentially based on the work of Chao (1959). The original subdivisions of Chao (1959), e.g. Flemingitan and Owenitan, are based on the scheme established by Spath (1934).

Although the application of Spath's subdivisions to the ammonoid succession of the Luolou Formation is correct in a global sense, it can be further refined. Tentative refinements by Chinese workers (see Tong & Yin 2002) did not include either systematic revision of the data provided by Chao, or detailed bed-by-bed sampling. The succession of Smithian ammonoids from Chaohu is reported to include *Flemingites–Euflemingites* Zone and the *Anasibirites* Zone (Tong *et al.* 2004). Very poor preservation may probably explain unusual associations such as *Owenites* or *Juvenites* with together prionitids (*Anasibirites* and *Wasatchites*), or the co-occurrence of *Anasibirites* with *Isculitoides*, the latter being typically restricted to the late Spathian.

This study presents for the first time, the distribution, based on bed-by-bed sampling, of Smithian ammonoids within the Luolou Formation. The zonation presented in this paper is partially new and greatly expands the number of successive palaeoequatorial ammonoid faunas (Fig. 15). No formal zone names are introduced here and we prefer to use the term: beds to describe the local faunal sequence. Ultimately, testing the validity and lateral reproducibility of formal zones should await biochronological analysis of data from other basins in conjunction with a fully standardized taxonomy.

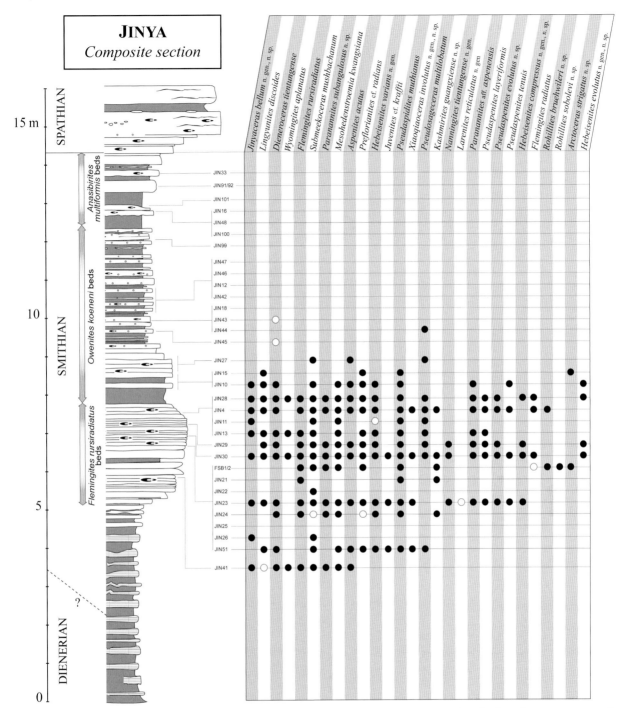

Fig. 3. Distribution of ammonoid taxa in the Jinya composite section. Open dots indicate occurrences based only on fragmentary or poorly preserved material.

Clypites *sp. indet. beds*

In Waili section, these beds that are characterized by the unique occurrence of *Clypites* sp. indet. and represent the youngest known Dienerian ammonoid fauna in the Luolou Formation. The general scarcity and incomplete knowledge of the late Dienerian ammonoids make it difficult to accurately place the Dienerian–Smithian boundary within the Luolou Formation.

Kashmirites kapila *beds*

This subdivision, representing a distinctive faunal association, coincides with the first occurrence of *Kashmirites*, which is associated with such diagnostic

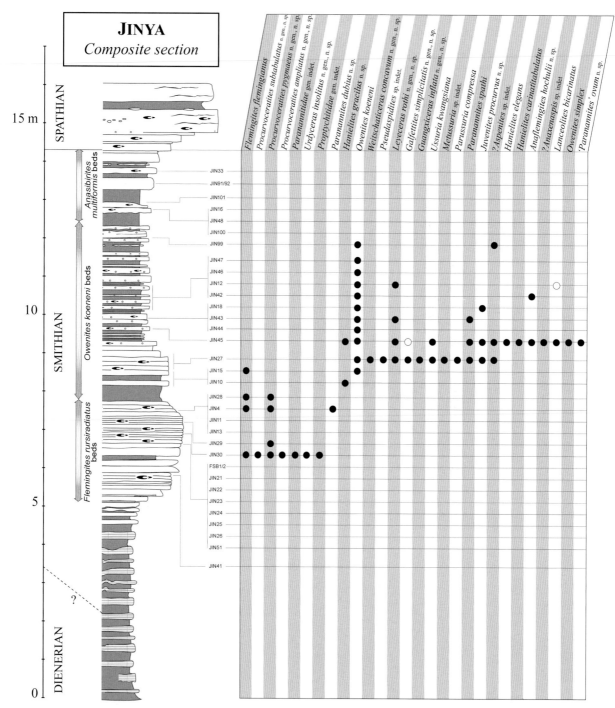

Fig. 4. Distribution of ammonoid taxa in the Jinya composite section. Open dots indicate occurrences based only on fragmentary or poorly preserved material.

genera as *Wailiceras*, *Guangxiceltites* or *Cordillerites*. Furthermore, the first occurrence of *Paranorites* and the last of '*Gyronites*' are also recorded in these beds. Typically, these beds occur just below the *Flemingites rursiradiatus* beds in a sequence of thin limestone beds and small-sized nodules. An exact correlative of this unique fauna does not exist within mid- and high-palaeolatitude basins. It has not yet been established whether the *Kashmirites kapila* beds is approximately equivalent to the upper part of the *H. hedenstroemi* Zone and the lower part of the *Euflemingites romunderi* Zone of mid-palaeolatitude sequences, or the lower part of the *Lepiskites kolymensis* Zone of the high-palaeolatitude record (Fig. 15).

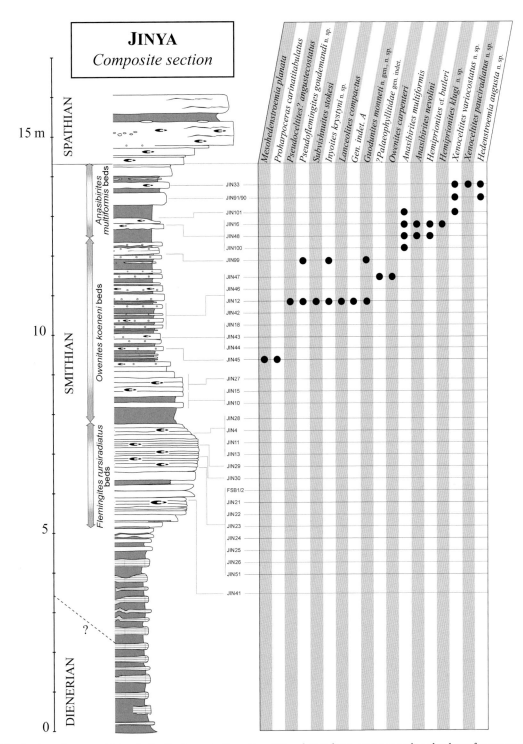

Fig. 5. Distribution of ammonoid taxa in the Jinya composite section. Open dots indicate occurrences based only on fragmentary or poorly preserved material.

Flemingites rursiradiatus *beds*

This subdivision is extremely well documented (Chao 1959; this work) and contains the most abundant, diverse Smithian fauna within the Luolou Fm. Among the most common genera are: *Flemin-* *gites*, *Rohillites*, *Pseudaspidites*, *Submeekoceras*, *Pseudaspenites* and *Mesohedenstroemia*. It is at least partly correlative with the *E. romunderi* Zone from British Columbia (Tozer 1994) and the *L. kolymensis* Zone from Siberia (Ermakova 2002).

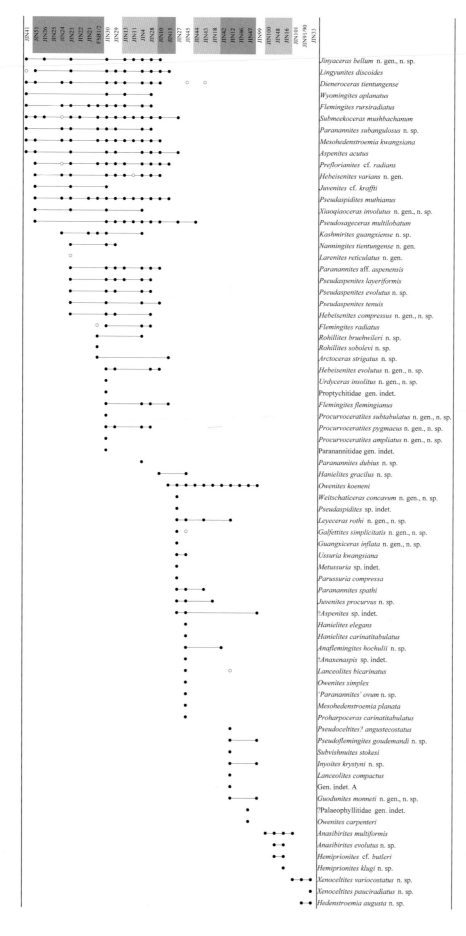

Fig. 6. Synthetic range chart of the succession of the Smithian ammonoid genera in the section of Jinya. Localities in grey indicate that they belong to the same group of beds (i.e. without superpositional information). Open dots indicate occurrences based only on fragmentary or poorly preserved material.

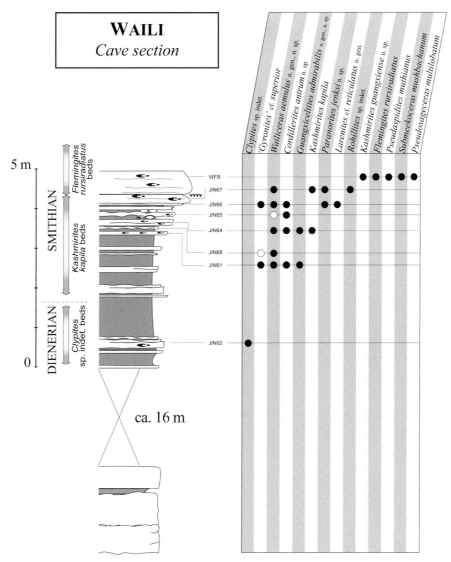

Fig. 7. Distribution of ammonoid taxa in the Waili Cave section. Open dots indicate occurrences based only on fragmentary or poorly preserved material.

Owenites koeneni *beds*

Although the *Owenites koeneni* beds include different successive faunas, *Owenites koeneni* is common to each of these. The succession is clearly displayed in Jinya and consists of the following horizons in ascending order:

- *Ussuria* horizon: characterized by the co-occurrence of *Ussuria kwangsiana* and *Metussuria* sp. indet.
- *Hanielites* horizon: characterized by the simultaneous but restricted presence of *Hanielites elegans*, *Proharpoceras carinatitabulatum* and '*Paranannites*' *ovum* n. sp. This subdivision is best documented in Yuping. It is usually restricted to a single bed in both Yuping and Jinya areas.

- *Inyoites* horizon: characterized by the co-occurrence of *Inyoites krystyni* n. sp. and *Pseudoflemingites goudemandi* n. sp.

With the exception of the long ranging and broadly distributed *Pseudosageceras*, the *Owenites koeneni* beds have no common species or genus with the mid- or high-palaeolatitude record, thus preventing recognition of exact correlatives. Indeed, the *Owenites koeneni* beds are composed of genera typically restricted to the intertropical belt.

Anasibirites multiformis *beds*

Diagnostic species occurring in this uppermost subdivision include *Anasibirites multiformis* and *Xenoceltites pauciradiatus* n. sp. This poorly diversified

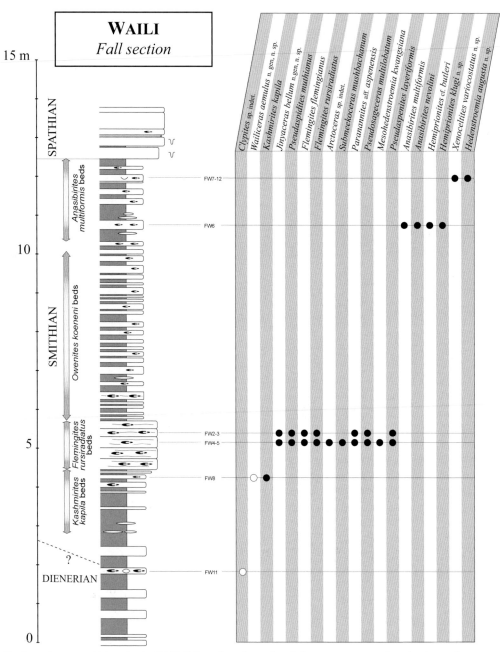

Fig. 8. Distribution of ammonoid taxa in the Waili Fall section. Open dots indicate occurrences based only on fragmentary or poorly preserved material.

fauna correlates with the *Anawasatchites tardus* Zone of British Columbia (Tozer 1994) and Siberia (Ermakova 2002).

Systematic palaeontology

All systematic descriptions follow the suprageneric classification established by Tozer (1981, 1994).

Biometric analyses follow the procedure of Monnet & Bucher (2005). Synonymy lists are annotated with the symbol system of open nomenclature recommended by Matthews (1973) and Bengtson (1988; v: specimens checked, p: reference applies only in part to the species under discussion, ?: allocation of the reference subject to some doubt, cf.: attribution to the species is possible but cannot be thought strictly certain).

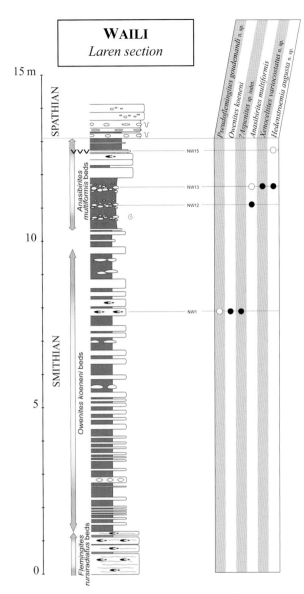

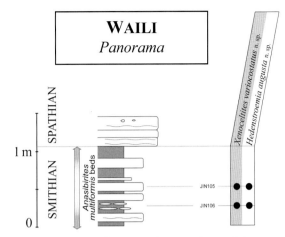

Fig. 10. Distribution of ammonoid taxa in the Waili Panorama section. Open dots indicate occurrences based only on fragmentary or poorly preserved material.

Fig. 9. Distribution of ammonoid taxa in the Waili Laren section. Open dots indicate occurrences based only on fragmentary or poorly preserved material.

Intraspecific variability: the population approach

Ammonoid species usually exhibit a continuous intraspecific variation ranging from involute, compressed and weakly ornamented variants to more evolute, depressed variants with coarser ornamentation. Within a single species, the frequency of these variants displays a typical normal distribution. This type of variation was coined 'First Buckman's Law of Covariation' by Westermann (1966) and has been well illustrated and discussed by e.g. Dagys & Weitschat (1993), Dagys *et al.* (1999), and Hammer & Bucher (2005).

Generally, robust variants present the most informative characters. Morphologies of smooth and compressed variants are less-well discernible and tend to converge across closely allied species or genera, thus making identification difficult. Thus, recognition of intraspecific variation is a crucial step for species identification. The statistical population approach usually has not been applied to Early Triassic ammonoids, with a few exceptions (e.g. Kummel & Steele 1962; Dagys & Ermakova 1988).

Interestingly, the morphological disparity of Smithian ammonoids seems to be restricted to a small number of morphological 'themes'. For example, the convergence of shell shapes is a frequent morphological phenomenon among Smithian ammonoids. The vast majority of shells lack marked ornamentation or are tabulate. Furthermore, the same type of ornamentation is repeated between different families. For instance, an extremely involute and oxycone shape is found in many genera such as *Hedenstroemia*, *Pseudosageceras* and *Cordillerites*. In the same way, many phylogenetically unrelated genera such as *Flemingites*, *Arctoceras* or *Dieneroceras* may be strigated. In most extreme cases, the suture line must be relied upon to provide key characters at the family level.

Measurements and statistical tests

The quantitative morphological range of each species is expressed utilizing the four classic geometrical parameters of the ammonoid shell: diameter (D), whorl height (H), whorl width (W) and umbilical diameter (U). However, whorl width measurement is often hindered by corrosion or dissolution of the upward side of the shell.

The three parameters (H, W and U) are plotted in absolute values as well as in relation to diameter

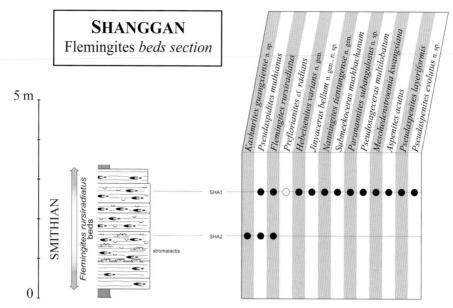

Fig. 11. Distribution of ammonoid taxa in the Shanggan section. Open dots indicate occurrences based only on fragmentary or poorly preserved material.

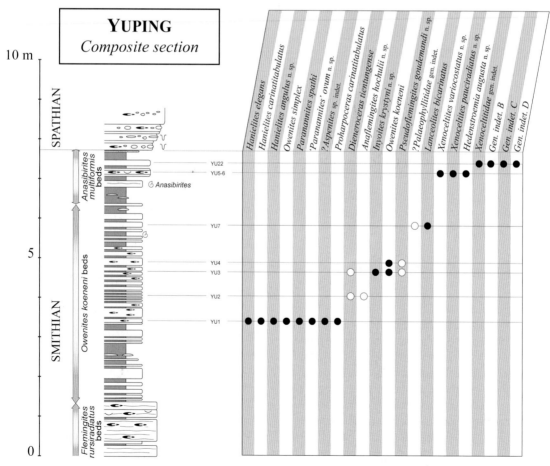

Fig. 12. Distribution of ammonoid taxa in the Yuping section. Open dots indicate occurrences based only on fragmentary or poorly preserved material.

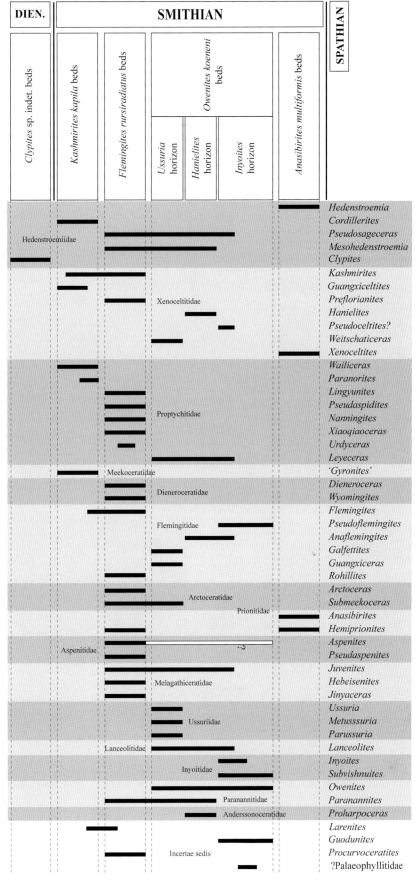

Fig. 13. Synthetic range chart showing the biostratigraphical distribution of Smithian ammonoid genera in northwestern Guangxi Province, South China (Dien. = Dienerian).

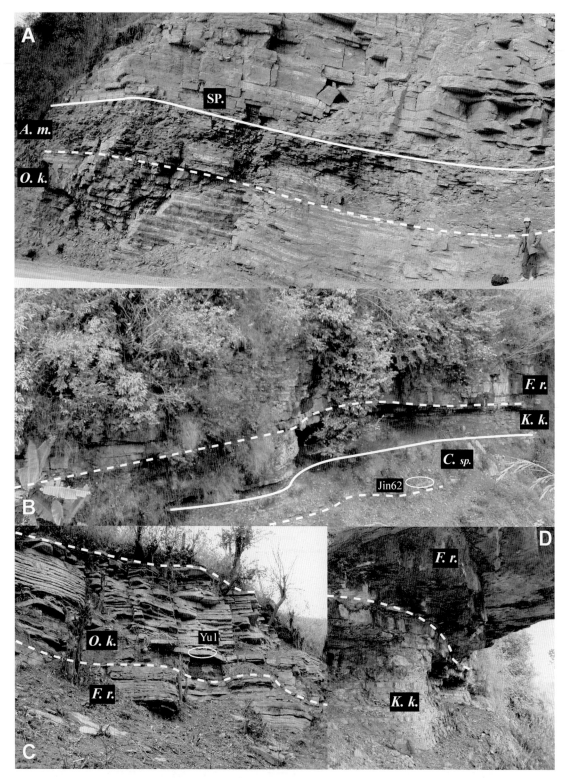

Fig. 14. □A. Laren section in Waili. Upper Smithian subdivisions are shown. Note the difference in facies between the Smithian and the Spathian beds. SP., Spathian; A. m., *Anasibirites multiformis* beds; O. k., *Owenites koeneni* beds. □B. Waili Cave section in Waili. Lower Smithian subdivisions are shown. F. r., *Flemingites rursiradiatus* beds; K. k., *Kashmirites kapila* beds; C. sp., *Clypites* sp. indet. beds. Bed denoted by black rectangle (Jin62) contains exclusively *Clypites* sp. indet. and represents the youngest Dienerian fauna within the Luolou Fm. Note the massive nature of the *Flemingites rursiradiatus* beds compared to the *Clypites* sp. indet. beds and *Kashmirites kapila* beds. □C. Yuping section. *Flemingites rursiradiatus* beds and *Owenites koeneni* beds are shown. Bed denoted by black rectangle (Yu1) contains *Proharpoceras*. □D. Waili Cave section in Waili. *Flemingites rursiradiatus* beds and *Kashmirites kapila* beds are shown.

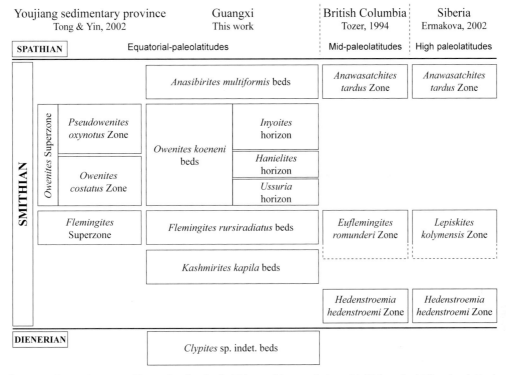

Fig. 15. Northwestern Guangxi ammonoid zonation for the Smithian and its correlation with high and middle palaeolatitude successions.

(H/D, W/D, and U/D). If a species is represented by at least 30 specimens, the normality of each parameter is graphically assessed by means of a probability plot (Monnet & Bucher 2005) and statistically tested by mean of a Lilliefors (1967) test. The Lilliefors test is a non-parametric 'closeness of fit' test of normality based on a correction of the Kolmogorov–Smirnov test for a sample with unspecified mean and variance (Lilliefors 1967). It evaluates the null hypothesis that the investigated data have a normal distribution with unspecified mean and variance, whereas the alternative hypothesis is that the investigated data do not have a normal distribution. The result of the test is indicated in the legend of the calculated normal curve associated with the measurements. A 'Normal' label indicates that the test cannot reject the null hypothesis of normality (at a confidence level of 95%), while 'Not normal' indicates that the hypothesis of a normal distribution is rejected (at a type I error rate < 5%). The normal probability plot presents a graphical test of normality. If the data conform to a normal distribution, the plot will be linear. Other probability density functions will generate departure from a linear plot.

Some species are also quantitatively compared by means of box and mean plots (Monnet & Bucher 2005). The box plot displays the 25th, 50th (median) and 75th percentiles of the range of measures covered by 99% of the specimens from a normally distributed sample. Outliers represent specimens not falling within the normal distribution. Furthermore, the mean plot displays the mean and its associated 95% confidence interval. Box and mean plots also allow for the graphical comparison of H, W and U for different species.

Allometry

Previous graphs and statistical tests have focused on the analysis of single parameters, which reflect only phenotypic differences. The growth trajectory of H, W and U (isometry or allometry) is also studied in order to detect and quantify possible heterochronic processes, or changes in size-based allometries of the geometry of the shell dimensional parameters.

Isometric growth implies that the parameter of interest has a constant ratio as a function of D, i.e. follows a linear equation. In contrast, allometric growth implies departure from linearity. To test the type of growth of each parameter, its values are fitted both by a linear and exponential equation by means of the reduced major axis fitting method (see Monnet & Bucher 2005). The resulting fitted curves are then tested by means of the coefficient of determination, the dispersion of residuals, and the Z-statistic associated with the allometric exponent. A better fit is obtained if the correlation coefficient tends towards one and the residuals tend to be closely scattered around the line. The Z-statistic tests the

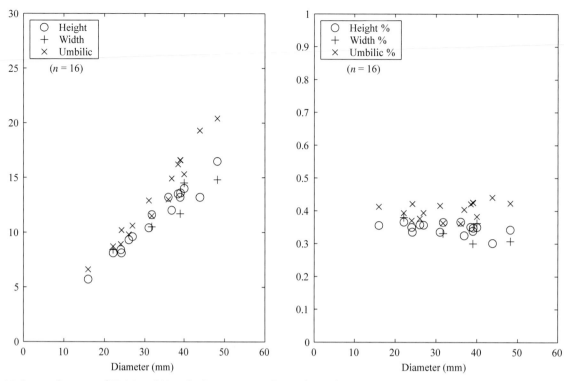

Fig. 16. Scatter diagrams of H, W, and U, and of H/D, W/D and U/D for *Kashmirites guangxiense* n. sp. (Jinya and Waili, *Flemingites rursiradiatus* beds).

null hypothesis that the allometric exponent is equal to one (i.e. isometric growth). In this study, allometry results are displayed as a graph representing the allometric or isometric growth trajectory of H, W and U.

Systematic descriptions

Repository of figured and measured specimens is abbreviated PIMUZ (Paläontologisches Institut und Museum der Universität Zurich). Locality numbers are reported on the measured sections (Figs 2–12, 14).

Class Cephalopoda Cuvier, 1797

Order Ammonoidea Zittel, 1884

Suborder Ceratitina Hyatt, 1884

Superfamily Xenodiscaceae Frech, 1902

Family Xenoceltitidae Spath, 1930

Genus *Kashmirites* Diener, 1913

Type species. – Celtites armatus Waagen, 1895

Kashmirites guangxiense n. sp.

Pl. 1: 1–10; Fig. 16

Occurrence. – Jin4, 21, 24, 30; FSB1/2; WFB; Sha2; *Flemingites rursiradiatus* beds.

Diagnosis. – Kashmirites with a subtabulate venter, rounded ventral shoulders, and swollen, distant, and fold-like radial ribs fading at maturity.

Holotype. – PIMUZ 25808, Loc. WSB, Waili, *Flemingites rursiradiatus* beds, Smithian.

Derivation of name. – Species name refers to the Guangxi Province.

Description. – Moderately evolute, thick platycone with a subtabulate to tabulate (for robust variant) venter, a rounded to subangular (for robust variant) ventral shoulder and nearly parallel to slightly convex flanks, forming a subquadratic whorl section. Umbilicus with moderately high, perpendicular wall and rounded shoulders. Ornamentation consists of distant, radial wavy ribs arising on umbilical shoulder and fading away on ventral shoulder. Ribs becoming somewhat bullate near mid-flank on immature stages. On some specimens, fine radial growth lines are visible on venter. Suture line ceratitic and simple with high ventral saddle and smaller umbilical saddle.

Discussion. – This species is characterized by a wide range of intraspecific variability. Robust variants have a thick quadratic section and subangular ventral shoulders. *K. guangxiense* n. sp. differs from the type species of the genus by having a quadratic whorl

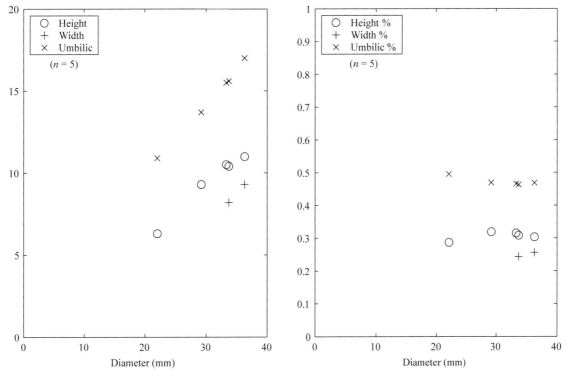

Fig. 17. Scatter diagrams of H, W, and U, and of H/D, W/D and U/D for *Kashmirites kapila* (Waili, *Kashmirites kapila* beds).

section, more distant ribs and fine radial growth lines and plications at maturity.

Kashmirites kapila (Diener, 1897)

Pl. 2: 1–4; Fig. 17

1897 *Danubites kapila* nov. sp. Diener, p. 50, pl. 15: 16a–c.

Occurrence. – Jin64, 67; FW8; *Kashmirites kapila* beds.

Description. – Very evolute, serpenticonic shell, with a low arched venter, rounded ventrolateral shoulders, and generally flat, parallel flanks, but slightly convex for compressed specimens. Umbilicus with moderately high, perpendicular wall and subangular shoulders. Closely spaced and pronounced ribs arising on umbilical shoulder, becoming weakly projected and fading on ventral shoulder. Suture line ceratitic with high ventral and lateral saddle. Umbilical saddle clearly smaller. Suture line of robust variants with broader saddles and lobes.

Discussion. – *K. densistriatus* (Welter, 1922), *K. evolutus* (Welter, 1922) and *K. kapila* all share a similar ribbing. However, *K. kapila* differs from the two other species by having a rectangular and laterally compressed whorl section. *K. kapila* is easily

distinguished from *K. guangxiense* n. sp. by its more evolute coiling, smaller whorl width, and finer ribbing (see Fig. 24). This species is used as an index for the *Kashmirites kapila* beds.

Genus *Preflorianites* Spath, 1930

Type species. – *Danubites strongi* Hyatt & Smith, 1905

Preflorianites cf. *radians* Chao, 1959

Pl. 2: 5–11; Fig. 18

1959 *Preflorianites radians* sp. nov. Chao, p. 196, pl. 3: 6–8.
1968 *Preflorianites* cf. *radians* – Zakharov, p. 137, pl. 27: 5–6.

Occurrence. – Jin4, 10, 13, 15, 23, 24, 28, 29, 30, 51; FSB1/2; *Flemingites rursiradiatus* beds.

Description. – Evolute, moderately compressed shell with a narrowly rounded to subangular venter (may vary from angular to subtabulate), rounded ventral shoulders, and parallel to slightly convex flanks. Umbilicus moderately high with perpendicular wall and rounded shoulders. Ornamentation consists of

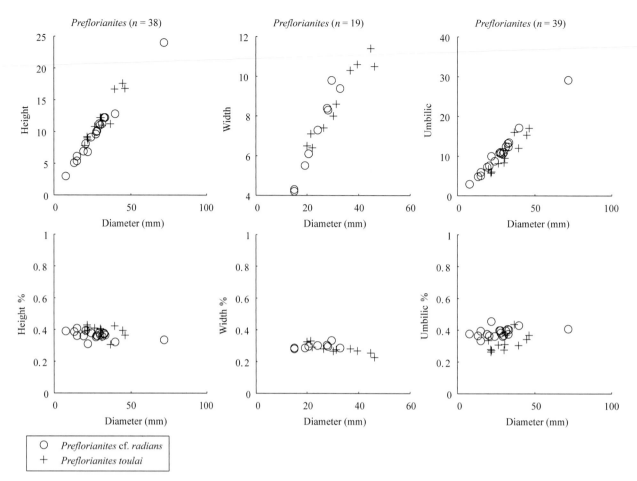

Fig. 18. Scatter diagrams of H, W, and U, and of H/D, W/D and U/D against corresponding diameter for *Preflorianites* cf. *radians* (*P. toulai* is given for comparison; data from Kummel & Steele 1962).

more or less dense, straight, radial ribs arising on umbilical shoulders and fading away low on ventral shoulders. Some very thin, radial, growth lines visible, especially on more compressed specimens. Venter completely smooth. Suture line ceratitic with serrated lobes, and ventral lobe more elongated than others.

Discussion. – This species essentially differs from *P. strongi* (Hyatt & Smith, 1905) by its subtabulate venter, and from *P. toulai* (Smith, 1932) by its more evolute coiling. The specimen referred to as *Preflorianites*? sp. by Tong *et al.* (2006, pl. 2: 15) is a flattened, smooth and involute ammonoid that must be excluded from this genus.

Genus *Pseudoceltites* Hyatt, 1900

Type species. – *Celtites multiplicatus* Waagen, 1895

Pseudoceltites? angustecostatus (Welter, 1922)

Pl. 3: 1–7; Fig. 19

?1922   *Xenodiscus angustecostatus* nov. sp. Welter, p. 110, pl. 4: 14–17.
?1922   *Xenodiscus oyensi* Welter, p. 111, pl. 5: 1, 2, 17.
?1968   *Anakashmirites angustecostatus* – Kummel & Erben, p. 128, pl. 19: 1–8.
?1973   *Anakashmirites angustecostatus* – Collignon, p. 144, pl. 5: 7–8.
v ?1978   *Eukashmirites angustecostatus* – Guex, pl. 7: 4.

Occurrence. – Jin12; T5; *Owenites koeneni* beds.

Description. – Slightly evolute, moderately compressed shell with a circular to broadly rounded venter, rounded ventral shoulders and convex flanks. Umbilicus with high, perpendicular wall and rounded shoulders. Ornamentation consists of dense, radial ribs arising on umbilical shoulder and ending high on ventral shoulders. Radial growth lines clearly visible, especially on ventral shoulder where ribs disappear. Suture line ceratitic with broad saddles, typical of Xenoceltitidae.

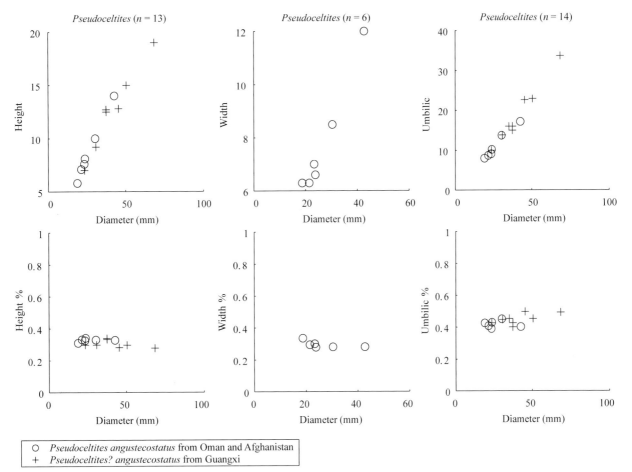

Fig. 19. Scatter diagrams of H, W, and D, and of H/D, W/D and U/D against corresponding diameter for *Pseudoceltites? angustecostatus* from Guangxi, and *P. angustecostatus* from Oman and Afghanistan (Oman and Afghanistan: data from Kummel & Erben 1968).

Discussion. – The morphology of this species is close to *Preflorianites* cf. *radians*, but it is slightly more evolute and has a more quadratic whorl section and more pronounced radial ribs, as well as visible growth lines (see Fig. 24). Its ribs are denser, more conspicuous and angular than those of *Preflorianites*. This species is assigned to *Pseudoceltites* (Hyatt, 1900) based on its type of ribbing. *P. angustecostatus* is also found associated with *Owenites* in Timor and Afghanistan. *Pseudoceltites? angustecostatus* from Guangxi is represented by a more inflated and rounded section than the type species of *Pseudoceltites*. This species is assigned to *Pseudoceltites* but it needs more data to strictly validate this attribution.

Genus *Hanielites* Welter, 1922

Type species. – *Hanielites elegans* Welter, 1922

Hanielites elegans Welter, 1922

Pl. 4: 1–5; Fig. 20

1922 *Hanielites elegans* nov. gen. et sp. Welter, p. 145, pl. 14: 7–11.

1934 *Hanielites elegans* – Spath, p. 243, fig. 82a–d.

v 1959 *Hanielites evolutus* sp. nov. Chao, p. 280, pl. 37: 8–12, text-fig. 36b.

v 1959 *Hanielites elegans* var. *involutus* var. nov. Chao, p. 281, pl. 37: 4–6, text-fig. 36a.

v 1959 *Hanielites rotulus* sp. nov. Chao, p. 281, pl. 37: 12–15.

v 1959 *Owenites kwangsiensis* sp. nov. Chao, p. 250, pl. 22: 1–2, 5–6.

Occurrence. – Jin45; Yu1; *Owenites koeneni* beds.

Description. – Small, slightly involute, somewhat thick platycone with a subangular to angular venter (sometimes bearing a delicate keel), rounded ventral shoulders, and flat, parallel flanks forming a generally quadratic whorl section. Shallow umbilicus with low, perpendicular wall and slightly rounded shoulders. Ornamentation is variocostate, varying from sinusoidal radiating plications to distinct ribs arising on umbilical

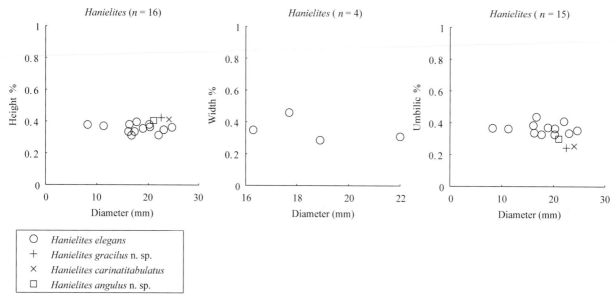

Fig. 20. Scatter diagrams of H/D, W/D and U/D against corresponding diameter for the different species of *Hanielites* (Jinya, *Owenites koeneni* beds).

shoulders, becoming forward projected low on ventral shoulder, and then crossing venter in a weak manner. Ribs become stronger approaching ventral shoulders and broken on ventral shoulders. Suture line ceratitic with broad lateral lobe.

Discussion. – This species is characteristic of the lowermost *Owenites koeneni* beds and is often found associated with *Proharpoceras*.

Hanielites carinatitabulatus Chao, 1959

Pl. 4: 9a–c; Fig. 20; Table 1

v 1959 *Hanielites carinatitabulatus* sp. nov. Chao, p. 282, pl. 37: 1–3, text-fig. 36c.

Occurrence. – Jin45; Yu1; *Owenites koeneni* beds.

Description. – Moderately involute, thick platycone with a subtabulate to broadly rounded venter (bearing delicate keel), rounded ventral shoulders and flat to slightly convex flanks forming a robust, quadratic whorl section. Umbilical wall higher than in *H. elegans*, but with oblique slope and rounded shoulders. Ornamentation consists of very distinct, prorsiradiate ribs arising on umbilical shoulder, developing great intensity on ventral shoulder, becoming strongly forward projected and rapidly disappearing near keel. No thin plications are visible. Ribs form weak tuberculation on inner whorls. Suture line ceratitic with saddles decreasing in size from venter to umbilicus.

Discussion. – This species is rare and is represented by a single specimen in Chao's collection and by only two specimens in our own material. *H. carinatitabulatus* differs from *H. elegans* by its more prominent ribs, absence of thin plications, its more involute coiling, and its deeper, narrower umbilicus.

Hanielites gracilus n. sp.

Pl. 4: 6–8; Fig. 20; Table 1

Occurrence. – Jin10, 45; *Owenites koeneni* beds.

Diagnosis. – *Hanielites* with a moderately involute, laterally compressed shell, a keeled, angular venter, and dense, but weak plications and prorsiradiate ribs.

Holotype. – PIMUZ 25834, Loc. Jin45, Jinya, *Owenites koeneni* beds, Smithian.

Derivation of name. – Species name refers to the alternation of small, thin ribs and plications.

Description. – Moderately involute, compressed shell with an angular venter bearing a distinctive keel, rounded ventral shoulders, and parallel flanks forming a subrectangular whorl section. Ornamentation consists of an alternation of thin prorsiradiate plications and ribs arising on flank above umbilical shoulder, then projecting forward onto venter and terminating near keel. Suture line ceratitic with broad, markedly indented lateral lobe. A small auxiliary series is also present.

Table 1. Measurements of some Smithian ammonoids from the Luolou Formation.

Genus	Species	Specimen number	D	H	W	U
Hanielites	*carinatitabulatus*	PIMUZ 25832	24	9.9	–	6.1
Hanielites	*gracile*	PIMUZ 25834	30.2	12.3	–	6.6
Hanielites	*angulus*	PIMUZ 25836	21	8.5	–	6.3
Weitschaticeras	*concavum*	PIMUZ 25869	40.5	11.8	–	19.7
Juvenites	cf. *kraffti*	PIMUZ 25914	13.6	3.6	8	6.1
Paranorites	*jenksi*	PIMUZ 25917	53	23.8	13.8	13
Paranorites	*jenksi*	PIMUZ 25919	10.6	4.7	3.2	3.3
Pseudaspidites	sp. indet.	PIMUZ 25935	22.3	11.5	–	4.4
Lingyunites	*tientungense*	PIMUZ 25944	23.4	13.4	6	–
Leyeceras	*rothi*	PIMUZ 25963	53.4	24.5	17.1	14.3
Urdyceras	*insolitus*	PIMUZ 25965	44	18.3	–	13.3
Proptychitidae gen. indet.		PIMUZ 25934	57.9	32.2	–	9.6
?*Anaxenaspis*	sp. indet.	PIMUZ 26015	12.3	35	–	40
Guangxiceras	*inflata*	PIMUZ 26016	134.1	50	26.8	52
Arctoceras	*strigatus*	PIMUZ 26024	39.9	19.8	–	7.7
Subvishnuites	*stokesi*	PIMUZ 26060	54.6	17.4	–	26.7
Paranannites	*dubius*	PIMUZ 26084	13.4	5.7	7.2	1.8
Paranannites	*dubius*	PIMUZ 26085	14.8	7	7.5	–
Paranannites	*dubius*	PIMUZ 26082	14	6.4	–	1.9
Paranannitidae gen. indet.		PIMUZ 26026	13	5.5	5.4	3.4
Owenites	*carpenteri*	PIMUZ 26191	23.5	13	–	–
Owenites	*carpenteri*	PIMUZ 26192	18.6	10.6	7.5	–
Mesohedenstroemia	*planata*	PIMUZ 26165	41	26.3	–	–
Guodunites	*monneti*	PIMUZ 26194	70.4	33.7	14.6	12.9
Procurvoceratites	*subtabulatus*	PIMUZ 26199	12.2	5.4	3.1	2.9
?Palaeophyllitidae gen. indet.		PIMUZ 25867	53.5	19.6	19.5	14.4

D, diameter; H, whorl height; W, whorl width; U, umbilical diameter.

Discussion. – This species is quite rare, but is easily distinguished from *H. elegans* by its more involute coiling, more angular ventral shoulders and its denser, more conspicuous alternation of ribs and plications. It also differs from *H. carinatitabulatus* by the presence of thin plications and a shallower umbilicus.

Hanielites angulus n. sp.

Pl. 4: 10a–d; Fig. 20; Table 1

Occurrence. – Yu1; *Owenites koeneni* beds.

Diagnosis. – Laterally compressed *Hanielites* with a subangular to narrowly rounded venter, rounded ventral shoulders, and weak, forward projecting, biconcave ribs disappearing near the venter.

Holotype. – PIMUZ 25836, Loc. Yu1, Yuping, *Owenites koeneni* beds, Smithian.

Derivation of name. – Species name refers to the angular shape of the venter.

Description. – Moderately involute, compressed platycone with a subangular to narrowly rounded venter (without keel), rounded ventral shoulders and parallel flanks. Umbilicus with low, perpendicular wall and rounded shoulders. Ornamentation consists of slightly prorsiradiate, biconcave ribs, thin and weak at umbilical shoulder, stronger on flank, disappearing on ventral shoulder. Suture line unknown.

Discussion. – This new species is represented by a single specimen. It is tentatively assigned to *Hanielites* because of its distinct morphology, but its peculiar ornamentation appears to be quite uncommon. However, the erection of a new genus cannot be justified on the basis of this single, fragmentary specimen.

Genus *Xenoceltites* Spath, 1930

Type species. – *Xenoceltites su*ᵇᵃᵛᵒˡᵘtus = *Xenodiscus* cf. *comptoni* (non Diener) Frebold, 1930

Xenoceltites variocostatus n. sp.

Pl. 5: 1–14; Pl. 6: 1–6; Fig. 21

?1895 *Dinarites coronatus* n. sp. Waagen, p. 27, pl. 7: 9–10.

Occurrence. – Jin33, 90, 91, 101, 105, 106; FW7, 12; NW13; Yu5, 6; *Anasibirites multiformis* beds.

Diagnosis. – Evolute *Xenoceltites* with extremely variocostate ribbing on inner whorls.

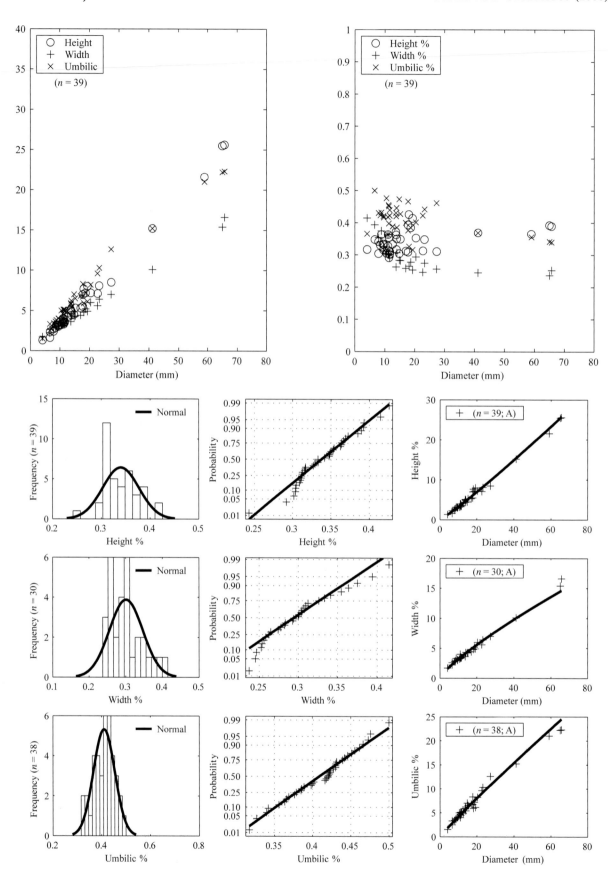

Fig. 21. Scatter diagrams of H, W, and U, and of H/D, W/D and U/D, and histograms, probability plots of H/D, W/D and U/D, and growth curves for *Xenoceltites variocostatus* n. sp. (Jinya and Yuping, *Anasibirites multiformis* beds). 'A' indicates allometric growth.

Holotype. – PIMUZ 25838, Loc. NW13, Waili, *Anasibirites multiformis* beds, Smithian.

Derivation of name. – Species name refers to its variocostate ribbing.

Description. – Slightly evolute, compressed platycone with rounded venter, rounded ventral shoulders, and slightly convex flanks, becoming gently convergent at a point low on ventral shoulder. Whorl section compressed in juvenile stages, becoming ovoid at maturity. Umbilicus wide, with moderately deep, oblique wall and rounded shoulders. For most juvenile specimens, ornamentation consists of conspicuous, sinuous, prorsiradiate ribs arising on umbilical shoulder, becoming strongly projected at ventral shoulders, and crossing venter with distinctive adoral curve. On some juvenile specimens, this strong forward projection and adoral curve imparts a somewhat crenulated appearance to the venter. These large, distinctive ribs gradually loose strength and eventually disappear in later stages. Growth lines are also visible and may be pronounced, particularly on adult whorls. Suture line ceratitic, typical of Xenoceltitidae, with broad ventral saddle and very small umbilical saddle.

Discussion. – Spath (1934) distinguished three species within the genus *Xenoceltites*: *X. subevolutus*, *X. spits-*

bergensis and *X. gregoryi*. These species essentially differ from each other according to their umbilical diameter and costation. The ornamentation on the new Chinese specimens invites comparison with *X. spitsbergensis* (e.g. bulges, ribs with a strong forward projection, and thin growth lines), but *X. spitsbergensis* has a more serpenticonic coiling. *X. subevolutus* seems to have about the same degree of involution as our specimens, but its ornamentation is less conspicuous and its whorl section is more compressed (see Weitschat & Lehmann 1978). Whorl height and width as well as umbilical diameter of *X. variocostatus* n. sp. exhibit significant allometric growth (Fig. 21). Height also tends to increase more rapidly than diameter for larger specimens. The measurements of *X. matheri* (Dagys & Ermakova 1990) are extremely close to our species, but it differs by its very weak ornamentation. Regardless of the motivation of various authors in differentiating between species, they are all morphologically similar and display wide ranges of intraspecific variation.

Xenoceltites pauciradiatus n. sp.

Pl. 6: 7–9; Fig. 22

?1978 *Xenoceltites subevolutus* Zakharov, pl. 11: 17.

Occurrence. – Jin33; Yu5; 6; *Anasibirites multiformis* beds.

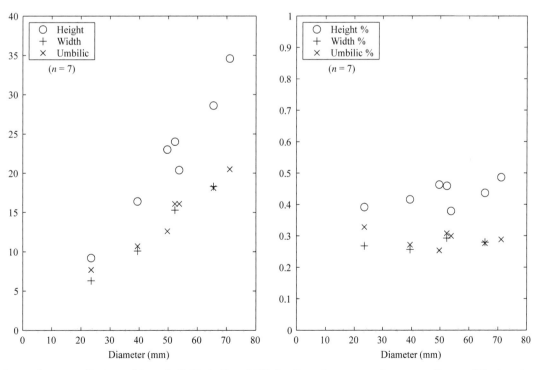

Fig. 22. Scatter diagrams of H, W, and U, and of H/D, W/D and U/D for *Xenoceltites pauciradiatus* n. sp. (Jinya and Yuping, *Anasibirites multiformis* beds).

Diagnosis. – Thick whorled *Xenoceltites* with moderately involute coiling. Variocostate ribbing restricted to the inner whorls.

Holotype. – PIMUZ 25858, Loc. Jin33, Jinya, *Anasibirites multiformis* beds, Smithian.

Derivation of name. – Species name refers to its weak ornamentation.

Description. – Moderately involute, compressed platycone with a highly arched venter, rounded ventral shoulders, and convex flanks with maximum curvature near umbilical shoulder. Umbilicus with moderately high, obliquely sloped wall and rounded shoulders. For juvenile specimens, ornamentation consists of very thin straight ribs that appear as weak tubercules on inner whorls. Mature specimens display fine, dense growth lines. Suture line unknown.

Discussion. – *X. pauciradiatus* n. sp. differs from *X. variocostatus* n. sp. by its more involute coiling, its greater whorl height in relation to diameter and its very weak ornamentation on adult specimens. Unfortunately, we did not find a sufficient number of specimens to statistically justify the erection of a new species. However, its phenotypic differences are hopefully characteristic.

Xenoceltitidae gen. indet.

Pl. 7: 8a–d

Occurrence. – Yu22; *Anasibirites multiformis* beds.

Description. – Evolute, compressed shell with a low rounded to subtabulate venter, rounded ventral shoulders and slightly convex flanks. Character of umbilicus difficult to determine on fragmentary specimen, but umbilical depth appears moderate with perpendicular wall and rounded shoulders. Ornamentation on body chamber consists only of distant plications. Very thin growth lines are also visible. Suture line ceratitic with three broad saddles.

Discussion. – The overall shape of this single specimen, its evolute coiling and suture line indicate that it probably belongs to Xenoceltitidae. It apparently differs from *X. variocostatus* n. sp. and *X. pauciradiatus* n. sp. by its more evolute coiling.

Genus *Guangxiceltites* n. gen.

Type species. – *Guangxiceltites admirabilis* n. sp.

Composition of the genus. – Type species only.

Diagnosis. – Small sized, moderately evolute xenoceltitid with rounded whorl section and forward projected ribs that cross the venter. Mature body chamber with growth striae only.

Derivation of name. – Genus name refers to the Guangxi Province.

Occurrence. – *Kashmirites kapila* beds.

Discussion. – This genus, with its rounded whorl shape and distinctive ribs that cross the venter, differs from all other xenoceltitids, which generally are more evolute and laterally compressed. This genus is too evolute to be included within Melagathiceratidae. *Guangxiceltites* n. gen. also differs from Melagathiceratidae by its projected ribs. The attribution of *Guangxiceltites* n. gen. to Xenoceltitidae must be confirmed by the discovery of new specimens exhibiting a well preserved suture line.

Guangxiceltites admirabilis n. sp.

Pl. 7: 1–6; Fig. 23

Diagnosis. – As for the genus.

Holotype. – PIMUZ 25861, Loc. Jin61, Waili, *Kashmirites kapila* beds, Smithian.

Derivation of name. – From the Latin word: admirabilis, meaning admirable.

Occurrence. – Jin61, 64; *Kashmirites kapila* beds.

Description. – Moderately evolute, slightly compressed shell with a broadly arched venter, rounded ventral shoulders, and nearly parallel flanks. Umbilicus with moderately high, oblique wall and rounded shoulders. Ornamentation consists of prorsiradiate ribs that weaken on the ventral shoulder before crossing the venter. Convex growth lines are also visible. Suture line unknown, all specimens completely recrystallized.

Discussion. – This species is morphologically close to *Juvenites*? *tenuicostatus* (Dagys & Ermakova, 1990), but it is thinner and more evolute. It differs from other xenoceltitids by its overall shape and smaller adult size, and by its ribs that tend to weaken near the ventral shoulder, but still cross the venter. However, its measurements exhibit the same overall percentages for height, width and diameter (see Fig. 24).

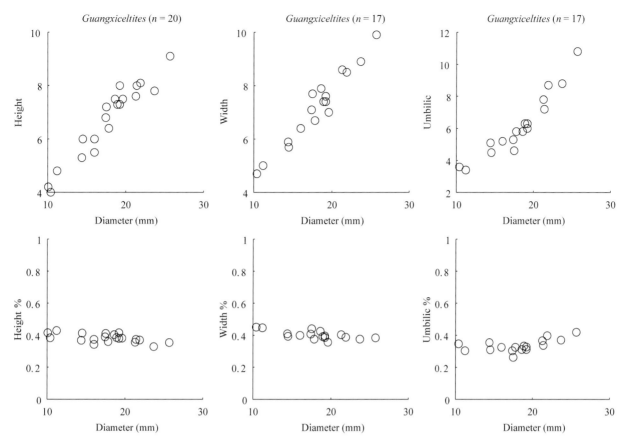

Fig. 23. Scatter diagrams of H, W, and U, and of H/D, W/D and U/D for *Guangxiceltites admirabilis* n. gen., n. sp. (Waili, *Kashmirites kapila* beds).

Genus *Weitschaticeras* n. gen.

Type species. – *Weitschaticeras concavum* n. sp.

Composition of the genus. – Type species only.

Diagnosis. – Laterally compressed, serpenticonic Xenoceltitidae with a tabulate venter and variable concave ribs.

Derivation of name. – Named after Wolfgang Weitschat (University of Hamburg, Germany).

Occurrence. – *Owenites koeneni* beds.

Discussion. – This new genus is easily distinguished by its conspicuous concave ribs, and its tabulate venter, which is relatively rare among the Xenoceltitidae.

Weitschaticeras concavum n. sp.

Pl. 7: 9a–d; Table 1

Diagnosis. – As for the genus.

Holotype. – PIMUZ 25869, Loc. Jin27, Jinya, *Owenites koeneni* beds, Smithian.

Derivation of name. – Species name refers to the characteristic concave ribs, from the Latin word: concavus.

Occurrence. – Jin27; *Owenites koeneni* beds.

Description. – Evolute serpenticone, with a tabulate venter, rounded ventral shoulders and parallel flanks. Shallow umbilicus with low, oblique wall and rounded shoulders. Ornamentation consists of variable, but distinct concave ribs that cross venter. Suture line ceratitic, typical for Xenoceltitidae, with broad lateral saddle.

Family Melagathiceratidae Tozer, 1971

Genus *Hebeisenites* n. gen.

Type species. – *Kashmirites varians* Chao, 1959

Composition of the genus. – Three species: *Kashmirites varians* Chao, 1959, *Hebeisenites evolutus* n. sp., and *Hebeisenites compressus* n. sp.

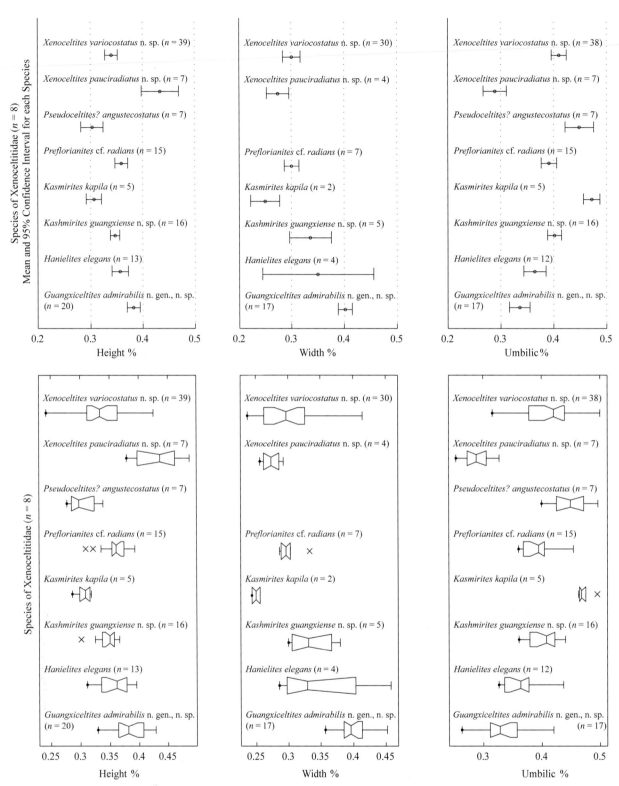

Fig. 24. Mean and box plots for different species of Xenoceltitidae found in the Luolou Formation.

Diagnosis. – Laterally compressed Melagathiceratidae with moderate to very evolute coiling, a laterally compressed whorl section, variable constrictions and a ceratitic suture line.

Derivation of name. – Named after Markus Hebeisen (Assistant, PIMUZ).

Occurrence. – *Flemingites rursiradiatus* beds and *Owenites koeneni* beds.

Discussion. – This new genus has a ceratitic suture line and its whorl section is completely different than the more globular *Thermalites*, and *Juvenites*-type species: *J. kraffti*. Thus, a new genus is erected based principally on its more compressed whorl section and simple ceratitic suture line. This new genus is very close to the laterally compressed *Prejuvenites* Waterhouse (1996a). The two genera could be congeneric. However, figured specimens of *Prejuvenites* are not well preserved and display a slightly more evolute coiling for largest specimens, with a whorl height superior to *Hebeisenites* n. gen. (Fig. 25; see measurements in Waterhouse 1996a). Constrictions are weaker in *Prejuvenites*. The latter also differs by a goniatitic suture line. Thus, it needs more data to strictly validate the congeneric hypothesis.

Hebeisenites varians (Chao, 1959)

Pl. 8: 1–11; Fig. 25

> 1959   *Kashmirites varians* sp. nov. Chao, p. 279, pl. 36: 1–6, 13; pl. 37: 18–19, 26–28.
> ?1959   *Kashmirites prosiradiatus* sp. nov. Chao, p. 278, pl. 36: 7; pl. 37: 29–31.
> ?1959   *Pseudoceltites contractus* sp. nov. Chao, p. 275, pl. 36: 8.
> ?1959   *Pseudoceltites ellepticus* sp. nov. Chao, p. 276, pl. 36: 9–12, 14–15, 30.
> ?1959   *Pseudoceltites kwangsianus* sp. nov. Chao, p. 276, pl. 36: 16–17.
> p ?1994   *Thermalites needhami* Tozer, p. 54, pl. 22: 5a–c only.

Occurrence. – Jin4, 11, 13, 23, 24, 28, 29, 30, 51; Sha1; T6, T50; *Flemingites rursiradiatus* beds. Jin10; *Owenites koeneni* beds.

Description. – Moderately evolute, small, compressed platycone with a subtabulate to rounded venter, subangular to abruptly rounded ventral shoulders, and parallel flanks gradually convergent to ventral shoulder, forming subrectangular to subquadratic whorl section. Shallow umbilicus with low, perpendicular wall and rounded shoulders. Ornamentation is characteristic for the species and consists of several different, but distinct constrictions that may be prorsiradiate and/or rursiradiate. Also, body chamber of mature specimens may exhibit small plications. These constrictions and plications cross the venter, but generally weaken before doing so. This species exhibits a very wide range of intraspecific variation with respect to height, width and ornamentation. Suture line weakly ceratitic and very simple, consisting of high ventral saddle and unique small, but wide lateral saddle.

Discussion. – Initially, this species was placed within the genus *Kashmirites* by Chao (1959). However, its very simple suture line and ornamentation consisting of distinct constrictions justify its assignment to the Melagathiceratidae. This genus represents an extreme variant of the family with its conspicuous lateral compression and moderately evolute coiling. The suture line initially illustrated by Chao appears to be quite peculiar and may be the result of excessive grinding. Whorl height of *Hebeisenites varians* exhibits isometric growth, whereas width and umbilical diameter display allometric growth (Fig. 25). The umbilicus tends to become proportionally more open as diameter increases.

Tozer (1994) illustrated a possible variant of *Thermalites needhami*, which is very much similar to *H. varians*, but he did not make a definitive assignment regarding the specimen, since its suture was not visible. Furthermore, this variant appears to be quite different from other specimens he assigned to this species (see pl. 22: 6a–b). Therefore, it probably should be assigned to *H. varians*.

Hebeisenites evolutus n. sp.

Pl. 8: 12–17; Fig. 26

Diagnosis. – *Hebeisenites* with very evolute coiling, an arched venter and deep constrictions.

Holotype. – PIMUZ 25886, Loc. Jin10, Jinya, *Flemingites rursiradiatus* beds, Smithian.

Derivation of name. – The species name refers to its evolute coiling.

Occurrence. – Jin28, 29, 30; *Flemingites rursiradiatus* beds. Jin10; *Owenites koeneni* beds.

Description. – Very evolute, moderately compressed, nearly serpenticonic shell with an arched venter, rounded ventral shoulders, and gently convex flanks. Umbilicus with perpendicular, moderately high wall and rounded shoulders. Ornamentation characteristic of Melagathiceratidae with deep, variable constrictions. Suture line possesses two different sized saddles, and appears to be goniatitic, but this may be due to excessive preparation, or the denticulations, if present, may be very few and very small.

Discussion. – This species can be distinguished from *H. varians* by its more evolute coiling (Fig. 25) and deeper constrictions.

Hebeisenites compressus n. sp.

Pl. 8: 18–25; Fig. 27

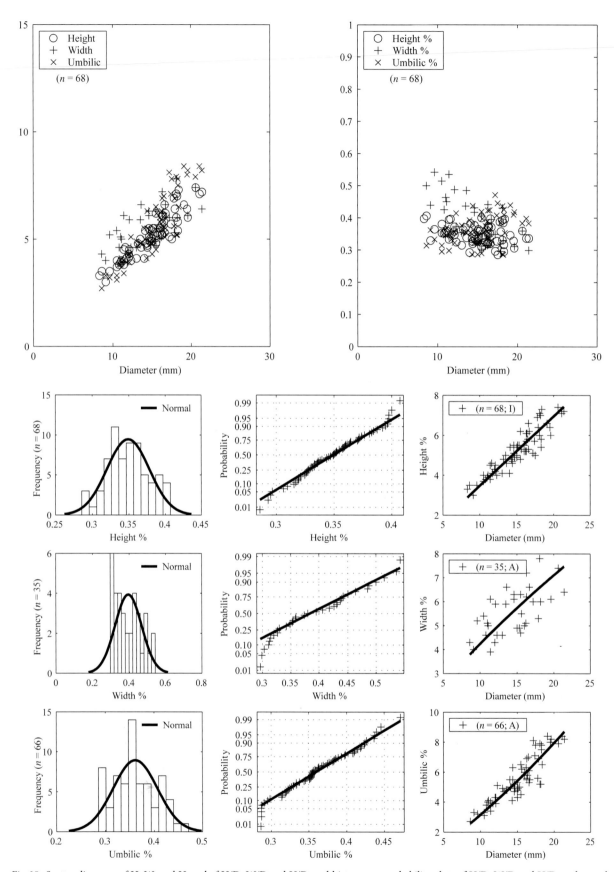

Fig. 25. Scatter diagrams of H, W, and U, and of H/D, W/D and U/D, and histograms, probability plots of H/D, W/D and U/D, and growth curves for *Hebeisenites varians* (Chao, 1959), (Jinya, *Flemingites rursiradiatus* beds). 'A' indicates allometric growth. 'I' indicates isometric growth.

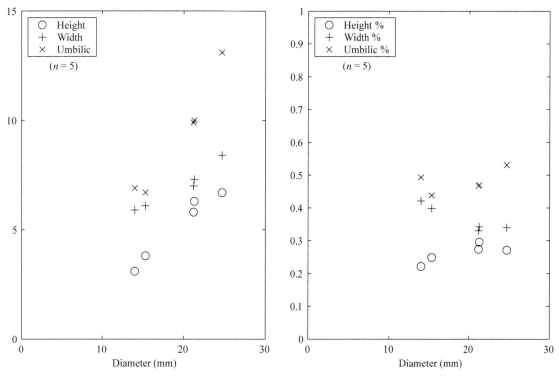

Fig. 26. Scatter diagrams of H, W, and U, and of H/D, W/D and U/D for *Hebeisenites evolutus* n. gen., n. sp. (Jinya, *Flemingites rursiradiatus* beds).

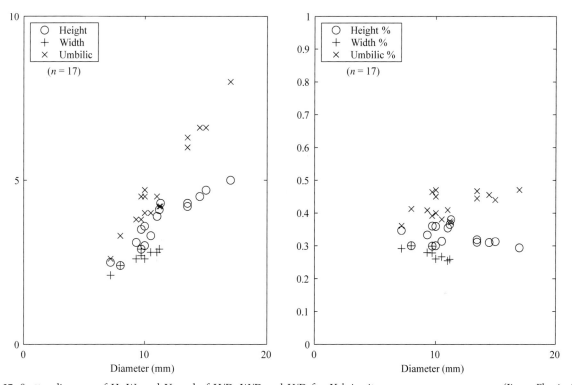

Fig. 27. Scatter diagrams of H, W, and U, and of H/D, W/D and U/D for *Hebeisenites compressus* n. gen., n. sp. (Jinya, *Flemingites rursiradiatus* beds).

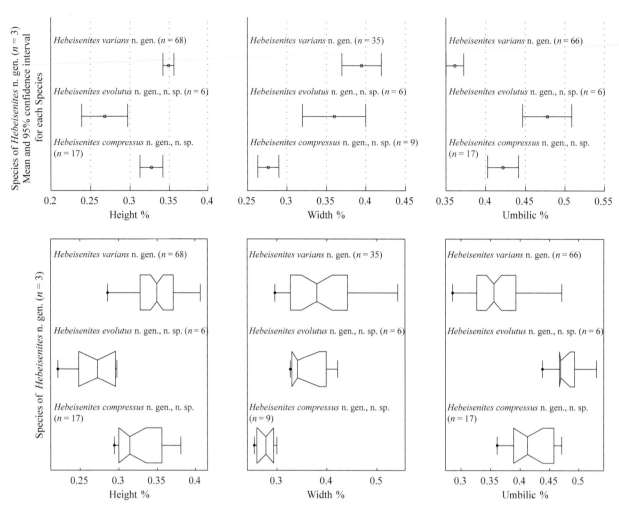

Fig. 28. Mean and box plots for different species of *Hebeisenites* found in the Luolou Formation.

Diagnosis. – Evolute, very compressed small-sized *Hebeisenites* with a tabulate venter, constrictions and a very simple suture line.

Holotype. – PIMUZ 25888, Loc. Jin30, Jinya, *Flemingites rursiradiatus* beds, Smithian.

Derivation of name. – Refers to its extreme lateral compression, from the Latin word: compressus.

Occurrence. – Jin23, 28, 29, 30; *Flemingites rursira-diatus* beds.

Description. – Evolute, very compressed, small shell with a tabulate to subtabulate venter, nearly angular ventral shoulders, and slightly convex flanks. Shallow umbilicus with oblique, low wall and rounded shoulders. Ornamentation consists of distinctive sinuous constrictions and plications. Suture line very

simple, with large lateral saddle and two large lobes. Lobes not well defined, but may be ceratitic.

Discussion. – This species can be distinguished from *H. varians* and *H. evolutus* n. sp. by its coiling, its more compressed shell, and its tabulate to subtabulate venter (Fig. 28).

Genus *Jinyaceras* n. gen.

Type species. – *Jinyaceras bellum* n. sp.

Composition of the genus. – Type species only.

Diagnosis. – Laterally compressed Melagathiceratidae with moderately evolute coiling, weakly convergent flanks, and variable, prorsiradiate constrictions.

Derivation of name. – Genus name refers to the small town of Jinya (Guangxi Province, South China).

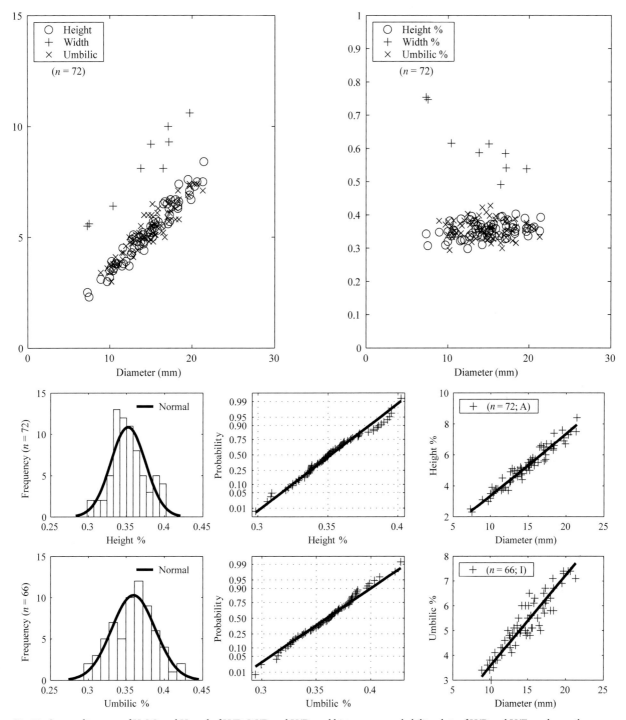

Fig. 29. Scatter diagrams of H, W, and U, and of H/D, W/D and U/D, and histograms, probability plots of H/D and U/D, and growth curves for *Jinyaceras bellum* n. gen., n. sp. (Jinya, *Flemingites rursiradiatus* beds). 'A' indicates allometric growth. 'I' indicates isometric growth.

Occurrence. – Flemingites rursiradiatus beds and *Owenites koeneni* beds.

Discussion. – Jinyaceras is distinguished from *Hebeisenites* by its more involute coiling, its more broadly arched venter, and its thicker whorls; from *Juvenites* by its more compressed shape.

Jinyaceras bellum n. sp.

Pl. 9: 1–19; Fig. 29

2007   Melagathiceratidae gen. et sp. nov. Klug *et al.*, text-fig. 4f–j.

Diagnosis. – As for the genus.

Holotype. – PIMUZ 25894, Loc. Jin28, Jinya, *Flemingites rursiradiatus* beds, Smithian.

Derivation of name. – From the Latin word: bellus, meaning handsome.

Occurrence. – Jin4, 11, 13, 23, 26, 28, 30, 41; FW2, 3, 4, 5; Sha1; T6, T50; *Flemingites rursiradiatus* beds. Jin10; *Owenites koeneni* beds.

Description. – Moderately evolute, subglobular shell with a broadly arched to nearly circular venter, rounded ventral shoulders and flanks varying from convex to broadly convergent from umbilical shoulder to circular venter. Umbilicus with moderately high, perpendicular wall and subangular shoulders. Umbilical width varies with amount of lateral compression. Ornamentation consists of variable, but distant prorsiradiate constrictions, which form very small, compact plications on body chamber of mature specimens. Constrictions nearly disappear on venter. Suture line simple and ceratitic, with a high and wide ventral saddle and smaller lateral saddle.

Discussion. – This species can be distinguished from *Hebeisenites varians* by its obviously more involute coiling and its conspicuously projected ornamentation. Whorl height is characterized by allometric growth, while umbilical diameter exhibits isometric growth (Fig. 29). Whorl height increases proportionally as diameter increases. *Jinyaceras bellum* n. sp. is closely allied to *Juvenites septentrionalis* Smith, 1932 by its similar ornamentation, but the latter's juvenile whorls are more involute. The suture line of juvenitids is also different (i.e. goniatitic).

Genus *Juvenites* Smith, 1927

Type species. – *Juvenites kraffti* Smith, 1927

Juvenites cf. kraffti Smith, 1927

Pl. 9: 20–23; Table 1

1927 *Juvenites kraffti* Smith, p. 23, pl. 21: 1–10.
1932 *Juvenites kraffti* – Smith, p. 109, pl. 21: 1–10.

Occurrence. – Jin23, 30, 51; *Flemingites rursiradiatus* beds.

Description. – Slightly evolute, subglobular *Juvenites* with an arched venter and a more or less depressed whorl section. Umbilicus with perpendicular, relatively high wall and rounded shoulders. Ornamentation consists of distant, forward projected constrictions. Suture line unknown for our specimens.

Discussion. – *J. kraffti* is distinguished by its very depressed whorl section and somewhat more evolute coiling. The scarcity of our specimens prevents us from utilizing statistics to establish their strict attribution to this species. Thus, we can only compare morphologically our specimens with *J. kraffti* and the species determination is uncertain.

Juvenites procurvus n. sp.

Pl. 22: 6–12; Fig. 30

?1959 *Juvenites septentrionalis* Chao, p. 289, pl. 25: 6–10.

Diagnosis. – *Juvenites* with dense, but distinct, straight constrictions projected towards the aperture.

Holotype. – PIMUZ 26010, Loc. T11, Tsoteng, *Owenites koeneni* beds, Smithian.

Derivation of name. – Species name refers to its forward projected constrictions, from the Latin words: pro- and -curvus meaning curved towards the aperture.

Occurrence. – Jin18, 27, 45; T5, 11, 12; Yu3; *Owenites koeneni* beds.

Description. – Moderately involute, globular shell with an arched venter and flanks convergent from umbilical shoulders to rounded venter (without distinct ventral shoulders). Umbilicus with a high, nearly perpendicular wall and somewhat abruptly rounded shoulders. Ornamentation consists of dense, straight, forward projected constrictions, becoming denser on mature body chamber. Constrictions appear more prominent near umbilicus. Suture line ceratitic with two broad saddles and a weakly indented lateral lobe.

Discussion. – This species is easily distinguished from other juvenitids by its straight, strongly forward, projected constrictions, which are denser on its body chamber than are those of *J. septentrionalis*. During preparation of the suture line, a weak indentation of the lateral lobe was revealed. Tozer (1981) divided the Melagathiceratidae according to the suture line (ceratitic or goniatitic). Many of the Melagathiceratidae illustrated in this study exhibit ceratitic suture lines, thus, contradicting Tozer's interpretation. Therefore, it is safe to assume that many genera of this family probably do not have a goniatitic suture line. However, *J. procurvus* is characterized by its globular shape and distinct constrictions, which are diagnostic of *Juvenites*.

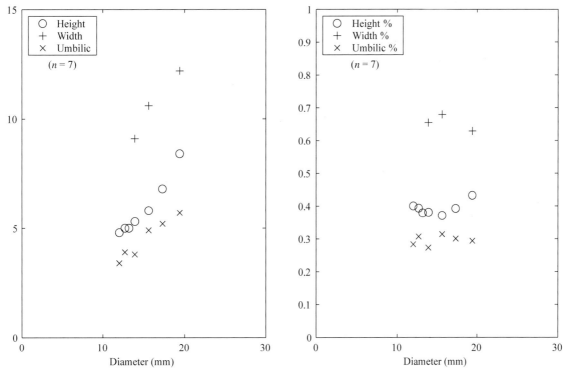

Fig. 30. Scatter diagrams of H, W, and U, and of H/D, W/D and U/D for *Juvenites procurvus* n. sp. (Jinya and Yuping, *Owenites koeneni* beds).

Superfamily Meekocerataceae Waagen, 1895

Family Proptychitidae Waagen, 1895

Genus *Paranorites* Waagen, 1895

Type species. – *Paranorites ambiensis* Waagen, 1895

Paranorites jenksi n. sp.

Pl. 9: 24–26; Table 1

?1959 *Paranorites ellipticus* sp. nov. Chao, p. 217, pl. 10: 9–10, text-fig. 16a.

Diagnosis. – Evolute proptychitid with slightly convex flanks, a subtabulate venter and sinuous plications near the umbilicus.

Holotype. – PIMUZ 25917, Loc. Jin67, Waili, *Kashmirites kapila* beds, Smithian.

Derivation of name. – Named after Jim Jenks (Salt Lake City).

Occurrence. – Jin66, 67; *Kashmirites kapila* beds.

Description. – Slightly involute, compressed platycone with a subtabulate venter, rounded ventral shoulders on mature specimens (slightly angular for juveniles), and nearly parallel flanks with weak curvature at mid-flank. Umbilicus with high, overhanging wall and narrowly rounded shoulders. Ornamentation consists of distant plications arising near umbilicus and disappearing on upper flanks. These plications are especially visible on juvenile specimens. Very thin growth lines are also visible. Ceratitic suture line typical of proptychitids with elongated, leaning saddles and indented lobes as well as a complex auxiliary series.

Discussion. – This species is very similar in shape to *Paranorites ambiensis* Waagen, 1895, but it clearly differs by its less complex suture line, which is actually closer to that of *Koninckites*. Morevover, *P. ambiensis* displays a distinct lateral saddle and is probably younger (Waagen 1895). This attribution to *Paranorites* is mainly based on the morphological similarity between *P. jenksi* n. sp. and the type species.

Genus *Pseudaspidites* Spath, 1934

Type species. – *Aspidites muthianus* Krafft & Diener, 1909

Pseudaspidites muthianus (**Krafft & Diener, 1909**)

Pl. 10: 1–10; Pl. 11: 1–4; Fig. 31

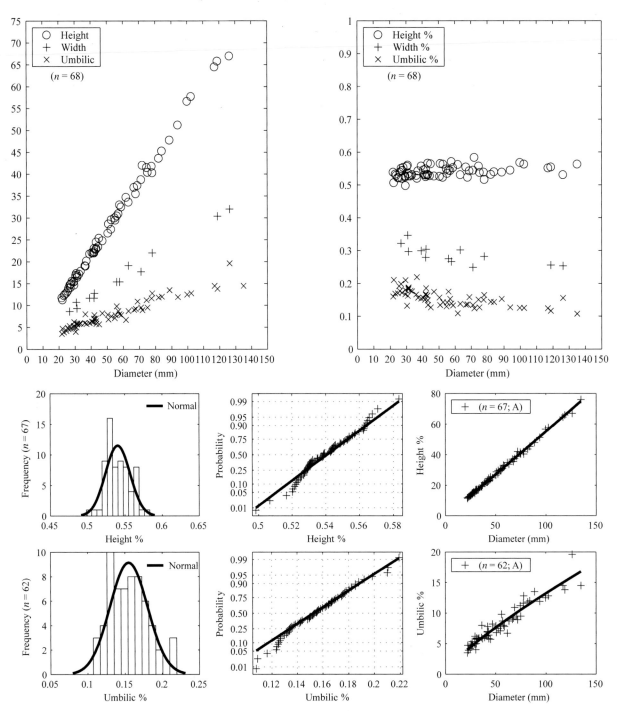

Fig. 31. Scatter diagrams of H, W, and U, and of H/D, W/D and U/D, and histograms, probability plots of H/D and U/D, and growth curves for *Pseudaspidites muthianus* (Jinya, *Flemingites rursiradiatus* beds). 'A' indicates allometric growth. 'I' indicates isometric growth.

1909 *Aspidites muthianus* Krafft & Diener, p. 59, pl. 6: 5; pl. 15: 1–2.

1932 *Clypeoceras muthianum* – Smith, p. 64, pl. 27: 1–7.

p 1932 *Ussuria waageni* Smith, p. 92 (*pars*), pl. 21: 34–36 only.

1934 *Pseudaspidites muthianus* – Spath, p. 164, fig. 48.

v 1959 *Pseudaspidites lolouensis* sp. nov. Chao, p. 229, pl. 13: 17–21, text-figs 20a, 21a.

v 1959 *Pseudaspidites kwangsianus* sp. nov. Chao, p. 230, pl. 12: 6–8, text-fig. 21d.

v 1959 *Pseudaspidites simplex* sp. nov. Chao, p. 231, pl. 13: 6–13; pl. 45: 5–7, text-figs 20b, 21b.

v 1959 *Pseudaspidites stenosellatus* sp. nov. Chao, p. 231, pl. 13: 4–5; pl.45:8, text-fig. 21c.

v 1959 *Pseudaspidites aberrans* sp. nov. Chao, p. 232, pl. 13: 14–15, text-fig. 20d.

v 1959 *Pseudaspidites longisellatus* sp. nov. Chao, p. 232, pl. 13: 1–3, text-fig. 20c.

?1959 *Proptychites pakungensis* sp. nov. Chao, p. 236, pl. 18: 1–2.

v 1959 *Proptychites hemialis* var. *involutus* Chao, p. 237, pl. 15: 13–16, text-fig. 24d.

?1959 *Proptychites markhami* Chao, p. 239, pl. 15: 3–5, text-fig. 23c.

v 1959 *Proptychites angusellatus* sp. nov. Chao, p. 240, pl. 15: 1–2.

v 1959 *Proptychites sinensis* sp. nov. Chao, p. 240, pl. 16: 5–6; pl. 17: 14–16, text-fig. 22c.

v 1959 *Proptychites latilobatus* sp. nov. Chao, p. 243, pl. 16: 1-2; pl. 19: 4–5.

v 1959 *Proptychites abnormalis* sp. nov. Chao, p. 243, pl. 16: 3–4.

v 1959 *Clypeoceras lenticulare* sp. nov. Chao, p. 225, pl. 12: 3–5, text-fig. 19b.

v 1959 *Clypeoceras tsotengense* sp. nov. Chao, p. 225, pl. 12: 1–2.

?1959 *Clypeoceras kwangiense* sp. nov. Chao, p. 226, pl. 17: 1–2, text-fig. 19a.

v 1959 *Ussuriceras*? sp. indet. Chao, p. 247, pl. 19: 1.

v 1959 *Pseudohedenstroemia magna* sp. nov. Chao, p. 265, pl. 41: 13–16; pl. 45: 1–2, text-fig. 32b.

?1962 *Pseudaspidites wheeleri* n. sp. Kummel & Steele, p. 673, pl. 101: 1, text-fig. 7c–e.

Occurrence. – Jin4, 11, 13, 15, 21, 23, 24, 28, 29, 30, 51; FSB1/2; WFB; FW2, 3, 4, 5; Sha1, 2; T6, T50; *Flemingites rursiradiatus* beds. Jin10; *Owenites koeneni* beds.

Description. – Very involute, compressed platycone with a subtabulate venter for mature specimens (rounded for juveniles), abruptly rounded to slightly angular ventral shoulders, and flat to slightly convex flanks. Narrow umbilicus, with high, perpendicular, wall and distinctive, abruptly rounded shoulder, similar to Arctoceratidae. Although most specimens are smooth with no discernable ornamentation, a few bear distant, straight or flexuous ribs. Umbilical bullae are present on only two specimens. When well-preserved, suture line exhibits distinctly phylloid saddles and well-individualized auxiliary lobe. Saddles tend to be curved adorally. If preparation is insufficient or preservation is poor, suture line can appear more or less complex (e.g. loss of the curved saddles; see pl. 10: 1d, 7e–10).

Discussion. – Preservation quality as well as laboratory preparation methods can directly affect the overall shape of these particular specimens, thus making it possible to confuse their identity with certain other genera, e.g. the Arctoceratidae. Likewise, the quality of preservation of this particular type of suture line can alter its appearance in such a manner that it could easily be mistaken as being representative of numerous different families ranging from the Ussuriidae to the Hedenstroemiidae.

Whorl height and umbilical diameter of *P. muthianus* exhibit allometric growth (Fig. 31). Estimated diameter of largest specimen exceeds 20 cm. *P. wheeleri* Kummel & Steele, 1962 essentially differs from *P. muthianus* by its suture line, which exhibits a greater individualization of the denticulations of the lobes. This difference may not be valid, and *P. wheeleri* may actually be a variant of *P. muthianus*. This hypothesis must be confirmed by sufficient measurements of *P. wheeleri*. Similarly, if other specimens with umbilical bullae are found, their morphological range must be determined to confirm whether the form is a variant.

Pseudaspidites sp. indet.

Pl. 11: 6a–d; Table 1

Occurrence. – Jin27; *Owenites koeneni* beds.

Description. – Morphologically similar to *P. muthianus*, but laterally compressed with nearly parallel flanks. Suture line similar to that of *P. muthianus*.

Discussion. – Since *Pseudaspidites* sp. indet. is represented by only one specimen, it cannot be assigned to *P. muthianus* with any degree of certainty.

Genus *Xiaoqiaoceras* n. gen.

Type species. – *Xiaoqiaoceras involutus* n. sp.

Composition of the genus. – Type species only.

Diagnosis. – Proptychitid with extremely involute coiling and a relatively simple suture line.

Derivation of name. – Named after Wan Xiaoqiao (China University of Geosciences, Beijing).

Occurrence. – *Flemingites rursiradiatus* beds.

Discussion. – The morphology of this genus is typical of proptychitids, except for its extremely involute coiling (occluded umbilicus). Compared to other

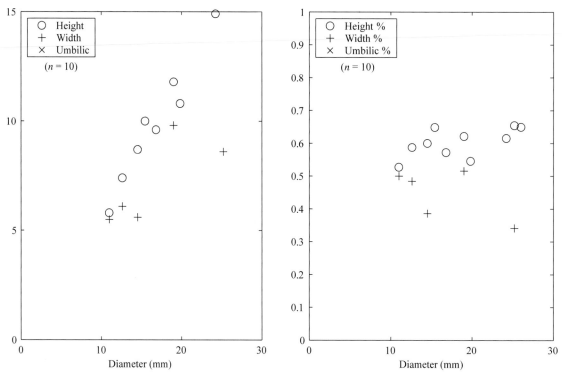

Fig. 32. Scatter diagrams of H, W, and U, and of H/D, W/D and U/D for *Xiaoqiaoceras involutus* n. gen., n. sp. (Jinya, *Flemingites rursiradiatus* beds).

proptychitid genera, its simplified suture line differentiates this genus within the family.

Xiaoqiaoceras involutus n. sp.

Pl. 12: 12–16; Fig. 32

Diagnosis. – As for the genus.

Holotype. – PIMUZ 25948, Loc. Jin4, Jinya, *Flemingites rursiradiatus* beds, Smithian.

Derivation of name. – Species name refers to its involute coiling.

Occurrence. – Jin4, 23, 30, 51; *Flemingites rursiradiatus* beds.

Description. – Extremely involute shell, with a subtabulate to rounded venter, rounded ventral shoulders, and slightly convex flanks. Umbilicus occluded, but rapid increase in whorl width forms distinctly rounded umbilical shoulders. Ornamentation smooth and only very thin growth lines are displayed on shell. Suture line is simplified compared to other proptychitids with small saddles, and with broadly denticulated lobes. Lateral lobe is broad,

trifid and deeply indented. An auxiliary series is present but less complex than, for instance in *P. muthianus*.

Discussion. – This species can be distinguished from other proptychitids by its extremely involute coiling and its peculiar, simplified suture line. However, as with all proptychitids, the suture line exhibits an obvious auxiliary series. This characteristic suture line may be a consequence of its extremely involute coiling and morphology. It can be noted that hedenstroemid genera also often display a trifid lateral lobe. However, the overall architecture of their suture lines is more complex.

Genus *Lingyunites* Chao, 1950

Type species: *Lingyunites discoides* Chao, 1950

Lingyunites discoides Chao, 1950

Pl. 12: 1–8; Fig. 33

1950 *Lingyunites discoides* gen. et sp. nov. Chao, p. 2, pl. 1: 1a–b, text-fig. 1.
1959 *Lingyunites discoides* – Chao, p. 223, pl. 11: 12–16, text-fig. 18.

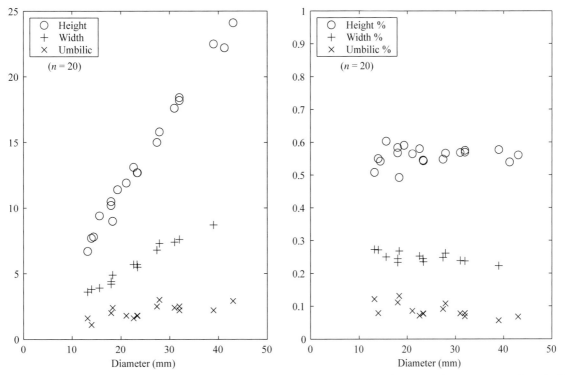

Fig. 33. Scatter diagrams of H, W, and U, and of H/D, W/D and U/D for *Lingyunites discoides* (Jinya, *Flemingites rursiradiatus* beds).

Occurrence. – Jin4, 13, 15, 23, 28, 29, 30, 41, 51; *Flemingites rursiradiatus* beds. Jin10; *Owenites koeneni* beds.

Description. – Very involute, discoidal shell, with a subtabulate to tabulate venter, abruptly rounded to slightly angular ventral shoulders, and nearly flat flanks with maximum curvature at mid-flank. Umbilicus very narrow, with moderately high, perpendicular wall and narrowly rounded shoulders. Ornamentation generally consists of very weak plications on flank near umbilicus, but a few small specimens exhibit sinuous plications across entire flank. Suture line ceratitic with three principal saddles and an individualized auxiliary series. Lateral saddle is turned towards the umbilicus and resembles suture line of *Pseudaspidites*.

Discussion. – This species can easily be confused with small specimens of *Mesohedenstroemia kwangsiana* due to its subtabulate venter. Furthermore, if the suture line is not well preserved, it can exhibit a similar, complex appearance. This genus, with its very involute coiling, is closely related to *Clypites*, differing mainly by its larger subtabulate to tabulate venter. Measurements reported by Chao (1959) correspond to ours, but the whorl width of Chao's specimens appears to be somewhat wider.

Genus *Nanningites* n. gen.

Type species. – *Clypeoceras tientungense* Chao, 1959

Composition of the genus. – Type species only.

Diagnosis. – Proptychitid with extremely involute coiling, a distinctive bicarinate venter and very angular ventral shoulders.

Derivation of name. – Genus name refers to the city of Nanning (Guangxi).

Occurrence. – *Flemingites rursiradiatus* beds.

Discussion. – The ornamentation of *Nanningites* closely resembles that of *Lingyunites discoides* Chao, 1950, but *N. tientungense* is clearly distinguished by its distinctive bicarinate venter and its nearly closed umbilicus.

***Nanningites tientungense* (Chao, 1959)**

Pl. 12: 9–11; Table 1

?1909 *Aspidites spitiensis* Krafft & Diener, p. 54, pl. 16: 3–8.

1959   *Clypeoceras tientungense* sp. nov. Chao,
       p. 228, pl. 17: 7–9.
1959   *Clypeoceras ensanumiforme* sp. nov. Chao,
       p. 227, pl. 17: 10–13, 16.

Occurrence. – Jin23, 29, 30; Sha1; T6, T50; *Flemingites rursiradiatus* beds.

Description. – Extremely involute, discoidal shell with a generally tabulate venter (bicarinate on some specimens), a nearly closed umbilicus, angular ventral shoulders, and weakly convex flanks. Umbilicus extremely narrow, with oblique, moderately high wall and abruptly rounded shoulder. Ornamentation consists of sinuous, forward projected plications, becoming more prominent on outer whorls. Fine growth lines parallel to plications occasionally visible. Ceratitic suture line appears less complex than that of *Lingyunites discoides*, but since it is not well preserved, considerable doubt exists regarding systematic affinities of this species.

Genus *Wailiceras* n. gen.

Type species. – *Wailiceras aemulus* n. sp.

Composition of the genus. – Type species only.

Diagnosis. – Smooth proptychitid with very involute, egressive coiling, and a tabulate venter.

Derivation of name. – The genus name refers to the Waili village (Guangxi Province, South China).

Occurrence. – *Kashmirites kapila* beds.

Discussion. – This genus exhibits strong affinities with Dienerian proptychitids such as *Koninckites*, but it also has its own distinctive characteristics, e.g. egressive coiling, a tabulate venter and a perpendicular umbilical wall. It also invites comparison with younger proptychitids such as *Pseudaspidites*, and thus, may represent a transitional morphology between Dienerian and Smithian proptychitids. The suture line has a well individualized auxiliary series, which indicates a close relationship with proptychitids rather with meekoceratids.

Wailiceras aemulus n. sp.

Pl. 13: 1–9; Fig. 34

?1909   *Meekoceras infrequens* Krafft & Diener, p. 44, pl. 30: 7a–d

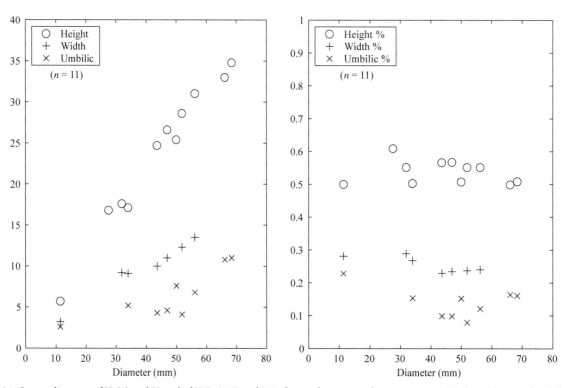

Fig. 34. Scatter diagrams of H, W, and U, and of H/D, W/D and U/D for *Wailiceras aemulus* n. gen., n. sp. (Waili, *Kashmirites kapila* beds).

Diagnosis. – As for the genus.

Holotype. – PIMUZ 25953, Loc. Jin61, Jinya, *Kashmirites kapila* beds, Smithian.

Derivation of name. – Name from the Latin word: *aemulus*, meaning imitating, and referring to its resemblance to the genus *Meekoceras*.

Occurrence. – Jin61, 64, 65, 67, 68; FW8; *Kashmirites kapila* beds.

Description. – Very involute, compressed, discoidal shell with a tabulate venter, very angular ventral shoulders, and slightly convex flanks having maximum curvature at mid-flank. Shell exhibits egressive coiling. Umbilicus relatively narrow, with high, perpendicular wall and abruptly rounded shoulders. Specimens are generally smooth, but larger sizes may exhibit very weak plications parallel to growth lines. Thin, weak, forward projected ribs are rarely seen on very small specimens. Ceratitic suture line, with well-developed auxiliary series, typical of proptychitids.

Discussion. – This species displays similarities with *Lingyunites discoides* Chao, 1950, such as a discoidal whorl section, a tabulate venter, and a similar suture line. However, measurements reported by Chao are different from ours.

   Wailiceras aemelus shows strong affinities with *Meekoceras infrequens* Krafft & Diener, 1909. Both apparently share very close shell shape. Their stratigraphical positions are compatible, and *M. infrequens* resembles a proptychitid more than a meekoceratid. The illustration of *M. infrequens* by Krafft & Diener (1909) is ambiguous, especially the rounded section of the last whorl. *W. aemulus* differs from *M. infrequens* by the presence of plications and a tabulate venter on the last whorl section. Since we have several specimens from Guangxi and a precise stratigraphical position for *W. aemulus*, it is preferable to create a species separate from *M. infrequens*.

Genus *Leyeceras* n. gen.

Type species. – *Leyeceras rothi* n. sp.

Composition of the genus. – Type species only.

Diagnosis. – Proptychitid with moderately evolute coiling, a subtabulate venter and thin lirae.

Derivation of name. – Genus name refers to the Leye city (Guangxi Province, South China).

Occurrence. – *Owenites koeneni* beds.

Discussion. – This proptychitid exhibits very close affinities with *Koninckites radiatus* Waagen, 1895. In fact, *K. radiatus* probably can be included within the variation of *L. rothi*, and since it does not agree well with the type species of *Koninckites*, we suggest it may actually be a synonym of *L. rothi*.

Leyeceras rothi n. sp.

Pl. 14: 1–3; Table 1

?1895 *Koninckites radiatus* n. gen. et sp. Waagen, p. 273, pl. 32: 2a–c.

Diagnosis. – As for the genus.

Holotype. – PIMUZ 25964, Loc. Jin43, Jinya, *Owenites koeneni* beds, Smithian.

Derivation of name. – Named after Rosi Roth (Preparator, PIMUZ).

Occurrence. – Jin12, 27, 43, 45; *Owenites koeneni* beds.

Description. – Moderately evolute shell with slightly convex flanks, a subtabulate venter, rounded ventral shoulders, and umbilicus with perpendicular, moderately high wall and rounded shoulders. Ornamentation consists of very weak, distant plications as well as radial lirae on flanks. Suture line typical of proptychitids with three principal elements slanted towards umbilicus. Although suture line on our specimen is not well preserved, it apparently has an indented ventral saddle.

Discussion. – In contrast to the illustrated specimen of *Koninckites radiatus* Waagen, 1895 (Waagen 1895, pl. 32: 2a–c), our species may exhibit a suture line with less numerous elements at a comparable diameter.

Genus *Urdyceras* n. gen.

Type species. – *Urdyceras insolitus* n. sp.

Composition of the genus. – Type species only.

Diagnosis. – Proptychitid with slightly involute coiling, a very tabulate venter and radial, distant folds.

Derivation of name. – Named for Séverine Urdy (Assistant, PIMUZ).

Occurrence. – Flemingites rursiradiatus beds.

Discussion. – This genus is similar in some features to *Meekoceras rota* Waagen, 1895, *Proptychites undatus* Waagen, 1895 and *Proptychites plicatus* Waagen, 1895. However, *P. undatus* and *P. plicatus* do not have a tabulate venter, and *M. rota* exhibits phylloid saddles.

Urdyceras insolitus n. sp.

Pl. 14: 4a–d; Table 1

Diagnosis. – As the genus diagnosis.

Holotype. – PIMUZ 25965, Loc. Jin30, Jinya, *Flemingites rursiradiatus* beds, Smithian

Derivation of name. – From the Latin word: insolitus, meaning uncommon.

Occurrence. – Jin30; *Flemingites rursiradiatus* beds.

Description. – Slightly involute shell with slightly convex flanks, a tabulate venter, angular ventral shoulders, and umbilicus with moderately high, perpendicular wall and rounded shoulders. Ornamentation consists of distant radial folds. Fine growth lines also visible. Suture line ceratitic with long, deeply indented lateral lobe, without auxiliary series. Lateral saddle gently inclined towards umbilicus.

Proptychitidae gen. indet.

Pl. 11: 5a–d; Table 1

Occurrence. – Jin30; *Flemingites rursiradiatus* beds.

Description. – Involute shell with a tabulate venter, rounded ventral shoulders, and convex flanks with maximum lateral curvature at mid-flank. Narrow umbilicus with moderately high, perpendicular wall and rounded shoulders. Ornamentation consists only of straight folds. Suture line peculiar with some similarity to proptychitids, particularly *Pseudaspidites*. Lobes appear less indented, but umbilical lobe seems to be curved. An auxiliary series is present, but is less divided. Most striking difference is absence of a ventral saddle. Although it may be present at larger diameters, our single specimen is not sufficiently well preserved to detect its presence.

Discussion. – With respect to shape and a portion of its suture line, our specimen exhibits some affinities with Proptychitidae. However, without additional specimens, we cannot assign it with any confidence to a known genus.

Family Meekoceratidae Waagen, 1895

Genus *Gyronites* Waagen, 1895

Type species. – Gyronites frequens Waagen, 1895

'*Gyronites*' cf. *superior* Waagen, 1895

Pl. 15: 1–3; Fig. 35

1895 *Gyronites superior* n. gen. et sp. Waagen, p. 294, pl. 37: 6a, b.

Occurrence. – Jin61, 66, 68; Kashmirites kapila beds.

Description. – Moderately evolute, very compressed shell with a tabulate venter, slightly angular ventral shoulders on adult specimens (very angular for juveniles), and convex flanks with maximum curvature at mid-flank. Mid-flank position of change in curvature is somewhat prominent, creating a very weak longitudinal ridge on flanks. Moderately wide, shallow umbilicus with very low, perpendicular wall and rounded shoulders. Ornamentation smooth with only strongly forward projected, fine, sinuous growth lines. Suture line typical for *Gyronites* with three elongated saddles, a deep umbilical lobe and a simple auxiliary series.

Discussion. – As noted by Waagen (1895), it is difficult to distinguish different species of *Gyronites* from the type species. Following the conclusions of Waagen, *G. superior* is differentiated by its slightly more involute coiling (see Waagen 1895, pl. 37: 1a, 6a) and its significantly more compressed whorl section. The flanks of *G. frequens* also appear to be more rounded. Our specimens are referable to *G. superior* even though small differences exist in the degree of lateral compression. To our knowledge, the material here attributed to *Gyronites superior* represents the youngest occurrence of this genus.

Family Dieneroceratidae Spath, 1952

Genus *Dieneroceras* Spath, 1934

Type species. – Ophiceras dieneri Hyatt & Smith, 1905

Dieneroceras tientungense Chao, 1959

Pl. 15: 5–12; Fig. 36

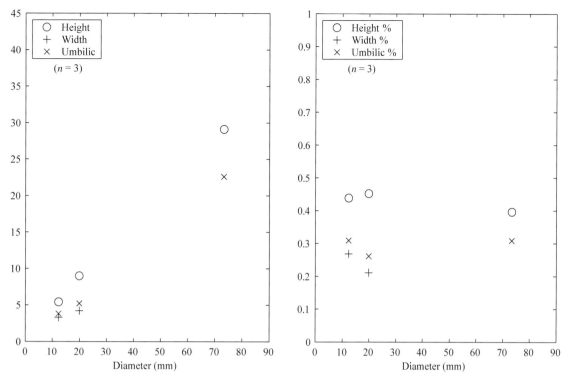

Fig. 35. Scatter diagrams of H, W, and U, and of H/D, W/D and U/D for 'Gyronites' cf. *superior* (Waili, *Kashmirites kapila* beds).

v 1959 *Dieneroceras tientungense* sp. nov. Chao, p. 192, pl. 2: 5–6, 8–10, 29.
v 1959 *Dieneroceras? vermiforme* sp. nov. Chao, p. 192, pl. 2: 14–16, 28, text-fig. 7b.
v 1959 *Dieneroceras ovale* sp. nov. Chao, p. 192, pl. 2: 11–13, text-fig. 7a.

Occurrence. – Jin4, 13, 23, 24, 28, 29, 30, 41, 43, 45, 51; Yu2, 3; *Flemingites rursiradiatus* beds. Jin10; *Owenites koeneni* beds.

Description. – Very evolute serpenticone exhibiting significant variation in whorl section, ranging from nearly ovoid with flat, parallel flanks and subtabulate venter, to gently rounded with slightly convex flanks and broadly arched venter. Ventral shoulders generally rounded. Very wide, fairly shallow umbilicus with moderately low, perpendicular wall and rounded shoulders. Ornamentation generally consists of very delicate strigation near venter, as well as a few weak folds and minor constrictions, especially on mature specimens. Suture line ceratitic, typical of family with two high ventral and lateral saddles and smaller umbilical saddle. Lobes are narrow.

Discussion. – Whorl height of *D. tientungense* is characterized by allometric growth, while umbilical diameter displays isometric growth (Fig. 36). Some

confusion surrounds this genus because it includes several species exhibiting a simple morphology, but with quite different rates of volution (e.g. *Dieneroceras spathi* Kummel & Steele, 1962 and *Dieneroceras knechti* (Hyatt & Smith, 1905)), and it also has been assumed to be a very long-ranging genus, first appearing in the Smithian and surviving until the Lower Spathian (Dagys & Konstantinov 1984). Furthermore, it has been synonymized with *Wyomingites* by Tozer (1981), thus creating confusion with regard to its classification. In addition, considerable disagreement exists with respect to its familial designation (e.g. Flemingitidae by Spath 1934 and Smith 1932; Dieneroceratidae by Kummel 1952 and Meekoceratidae by Tozer 1981). Its rate of coiling and ornamentation (especially the strigation) closely mimic the Flemingitidae and suggest possible links with this family. Yet, its suture line is simpler and quite different. Based on its morphology, it is difficult to understand the justification for assigning it to a different family, and especially the Meekoceratidae.

Spathian species attributed to this genus have different ornamentation (e.g. tubercules: *D. tuberculatum* Dagys & Konstantinov, 1984) and often, a more complex suture line (*D. demokidovi* Dagys & Konstantinov, 1984). Their assignment to *Dieneroceras* remains doubtful.

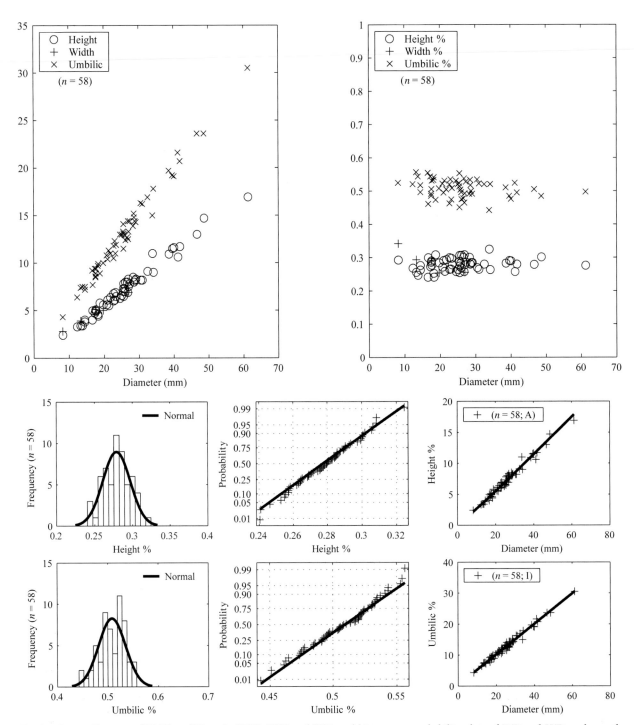

Fig. 36. Scatter diagrams of H, W, and U, and of H/D, W/D and U/D, and histograms, probability plots of H/D and U/D, and growth curves for *Dieneroceras tientungense* (Jinya, *Flemingites rursiradiatus* beds). 'A' indicates allometric growth. 'I' indicates isometric growth.

Genus *Wyomingites* Hyatt, 1900

Type species: *Meekoceras aplanatum* White, 1879

***Wyomingites aplanatus* (White, 1879)**

Pl. 16: 1–3; Fig. 37

1879   *Meekoceras aplanatum* White, p. 112.
1880   *Meekoceras aplanatum* – White, p. 112, pl. 31: 1a–b, d.
1900   *Wyomingites aplanatus* – Hyatt, p. 556.
1902   *Ophiceras aplanatum* – Frech, p. 631.
1905   *Meekoceras (Gyronites) aplanatum* – Hyatt & Smith, p. 146, pl. 11: 1–14; pl. 64: 17–22; pl. 77: 1–2.

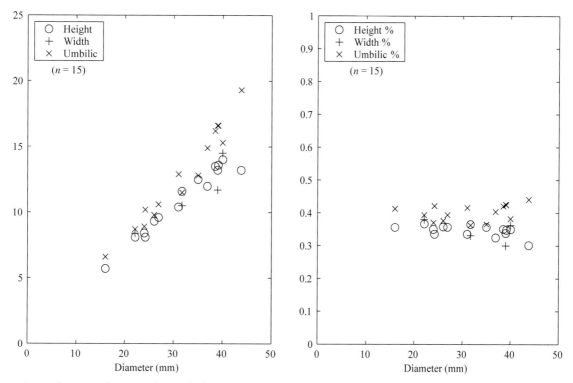

Fig. 37. Scatter diagrams of H, W, and U, and of H/D, W/D and U/D for *Wyomingites aplanatus* (Jinya, *Flemingites rursiradiatus* beds).

1932 *Flemingites aplanatus* – Smith, p. 51, pl. 11: 1–14; pl. 22: 1–23; pl. 39: 1–2; pl. 64: 17–32.
1962 *Wyomingites* cf. *aplanatus* – Kummel & Steele, p. 696, pl. 99: 3–4.
?1979 *Wyomingites aplanatus* – Nichols & Silberling, pl. 1: 19–21.

Occurrence. – Jin13, 28, 30, 41; *Flemingites rursiradiatus* beds.

Description. – Very evolute, laterally compressed shell with a subrectangular whorl section, a tabulate venter, rounded ventral shoulders, and nearly parallel flanks. Moderately wide umbilicus with low, perpendicular wall and rounded shoulders. Ornamentation consists of conspicuous strigation (as in *Flemingites*) and highly variable, dense to distant wavy ribs, varying also in strength. Ribs may be replaced by small plications at maturity. Suture line ceratitic, simple and structurally similar to *Dieneroceras*.

Discussion. – The strigate ornamentation of this genus can easily cause it to be confused with flemingitids, and yet, its suture line is similar to that of *Dieneroceras*. Tozer (1981) synonymized *Dieneroceras* with *Wyomingites*, which he then placed within the Meekoceratidae. However, their respective morphology and suture line are similar enough to justify placing these two genera in a specific family: the Dienerocer-

atidae. Compared to *W. scapulatus* Tozer, 1994, *W. aplanatus* is more evolute and displays more prominent strigation as well as more variable ribbing. This genus apparently exhibits highly variable ornamental strength.

Family Flemingitidae Hyatt, 1900

Genus *Flemingites* Waagen, 1892

Type species. – *Ceratites flemingianus* de Koninck, 1863

Flemingites flemingianus (de Koninck, 1863)

Pl. 17: 1–5; Fig. 38

1863 *Ceratites flemingianus* de Koninck, p. 10, pl. 7: 1.
1895 *Flemingites flemingianus* de Koninck sp. – Waagen, p. 199, pl. 12: 1; pl. 13: 1; pl. 14: 1.
1933 *Flemingites flemingi* var. *madagascariensis* nov. var. Collignon, p. 25, pl. 5: 1, fig. 9.
v 1959 *Flemingites ellipticus* sp. nov. – Chao, p. 206, pl. 4: 5–7, 10–12, text-fig. 12a.

Occurrence. – Jin4, 15, 28, 30; FW4/5, FW2, 3, 4, 5; T6, T50; *Flemingites rursiradiatus* beds.

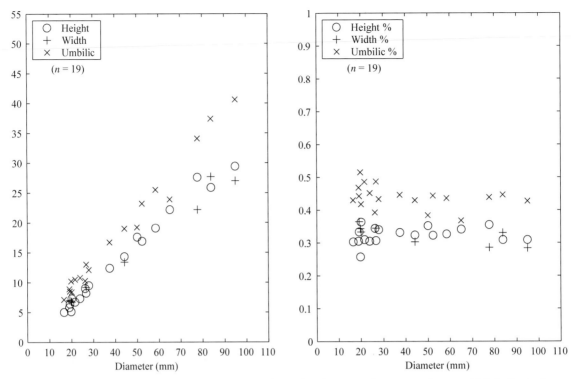

Fig. 38. Scatter diagrams of H, W, and U, and of H/D, W/D and U/D for *Flemingites flemingianus* (Jinya and Waili, *Flemingites rursiradiatus* beds).

Description. – Evolute shell exhibiting a subcircular to subquadratic whorl section with a broadly rounded to circular venter and slightly convex flanks, gently converging towards venter. Umbilicus moderately wide with high, perpendicular wall and broadly rounded shoulders. Ornamentation consists of noticeable radial or slightly rursiradiate ribs, as well as very conspicuous, dense strigation covering entire shell at all growth stages larger than ca. 1 cm in diameter. Suture line ceratitic with well-indented lobes and subphylloid saddles.

Discussion. – *Flemingites compressus* Waagen, 1895, from the Salt Range, with its more ovoid whorl section may represent a variant of the type species. The suture line of *F. flemingianus* is apparently highly variable.

Flemingites rursiradiatus Chao, 1959

Pl. 18: 1–7; Pl. 19: 1–3; Fig. 39

v 1959   *Flemingites rursiradiatus* sp. nov. Chao, p. 205,
              pl. 6: 1–5, 8–10, text-fig. 13a–b.
p ?1933 *Flemingites griesbachi* Collignon, p. 28 (*pars*),
              pl. 6: 2, 2a only.

Occurrence. – Jin4, 13, 21, 23, 24, 28, 29, 30, 41; FSB1/2; WFB; FW2, 3, 4, 5; Sha1, 2; *Flemingites rursiradiatus* beds.

Description. – Laterally compressed serpenticone with a subtabulate venter, rounded ventral shoulders, and subrectangular whorl section for robust specimens, and a narrower, tabulate venter with subangular ventral shoulders on more compressed specimens. Flanks parallel near umbilicus, then gradually converge to the venter. Fairly wide umbilicus with moderately high, perpendicular wall and rounded shoulders. Ornamentation consists of strigation on flanks and variable strength, conspicuous rursiradiate ribs, arising on umbilical shoulder, usually becoming very faint on ventral shoulder, then crossing venter in a manner ranging from barely perceptible to highly conspicuous. These ribs may become straight to slightly concave on adult specimens, and the intensity of more robust ribs can create a 'polygonal coiling' effect as they cross the venter. Suture line ceratitic with rounded (but not completely phylloid) lateral saddle.

Discussion. – This species differs from other flemingitids by its conspicuous rursiradiate ribs on the inner whorls, which disappear on the venter. Whorl height and umbilical diameter exhibit isometric growth (Fig. 39). Apparently, this species displays an extremely wide range of intraspecific variation, and the loss of rursiradiate ribs on some adult specimens raises the possibility that many of the larger specimens (e.g. several of Diener's Himalayan specimens)

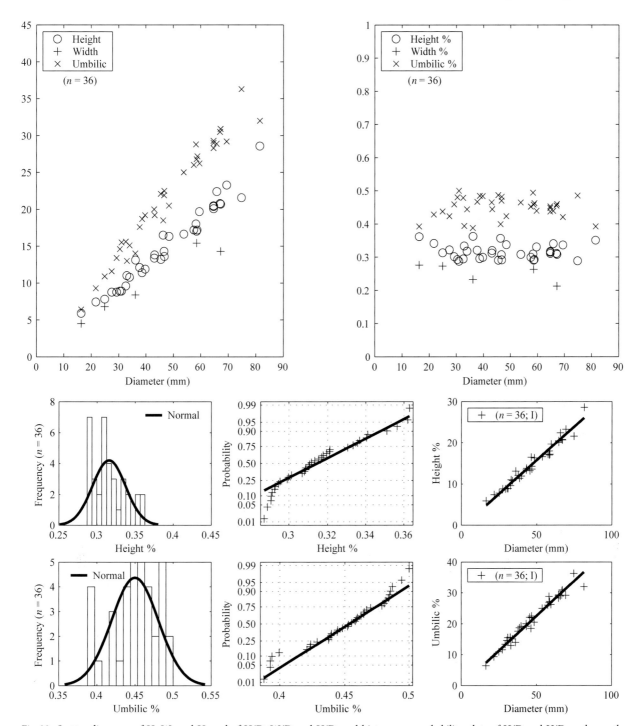

Fig. 39. Scatter diagrams of H, W, and U, and of H/D, W/D and U/D, and histograms, probability plots of H/D and U/D, and growth curves for *Flemingites rursiradiatus* (Jinya, *Flemingites rursiradiatus* beds). 'A' indicates allometric growth. 'I' indicates isometric growth.

originally assigned to different species may, in fact, belong to *F. rursiradiatus*.

Flemingites radiatus Waagen, 1895

Pl. 19: 4–6; Fig. 40

1895   *Flemingites radiatus* n. gen. et sp. Waagen, p. 197, pl. 11: 1a–b.

?1933   *Flemingites radiatus* – Collignon, p. 24, pl. 5: 2–3.

1934   *Flemingites radiatus* – Spath, p. 111, fig. 28.

Occurrence. – Jin4, 28, 30; FSB1/2; T6, T50; *Flemingites rursiradiatus* beds.

Description. – Moderately evolute, laterally compressed shell with a rectangular whorl section, parallel flanks,

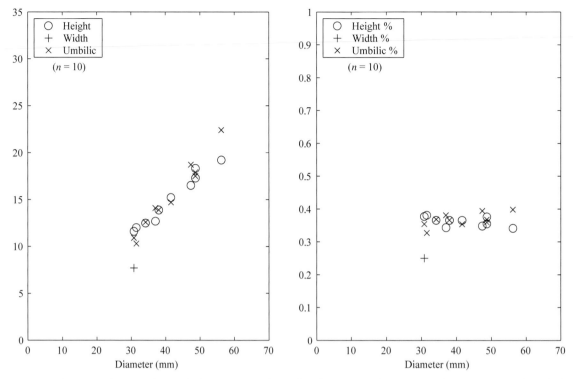

Fig. 40. Scatter diagrams of H, W and U, and of H/D, W/D and U/D for *Flemingites radiatus* (Jinya, *Flemingites rursiradiatus* beds).

a broadly arched to subtabulate venter and rounded ventral shoulders. Umbilicus wide with moderately high, perpendicular wall and rounded shoulders. Ornamentation consists of forward projected, weak ribs and weak strigation on flanks. Suture line ceratitic and similar to *Flemingites rursiradiatus*.

Discussion. – F. radiatus is characterized by its conspicuous rectangular section, and its coiling is more involute than *F. rursiradiatus*. This species can be confused with *Wyomingites aplanatus* since they both have laterally compressed shells with (sub)rectangular whorl sections and similar strigate ornamentation. The much weaker plications and the less rectangular whorl section of *F. radiatus* provide means of distinction with *W. aplanatus*.

Genus *Rohillites* Waterhouse, 1996b

Type species. – Flemingites rohilla Diener, 1897

Rohillites bruehwileri n. sp.

Pl. 20: 3–8; Fig. 41

Diagnosis. – Moderately evolute *Rohillites* with pronounced and radiate ribs and plications. Strigation restricted to lower flanks and venter.

Holotype. – PIMUZ 26478, Loc. Jin4, Jinya, *Flemingites rursiradiatus* beds, Smithian.

Derivation of name. – Species named for Thomas Brühwiler (Assistant, PIMUZ).

Occurrence. – Jin4; FSB1/2; *Flemingites rursiradiatus* beds.

Description. – Evolute and laterally compressed shell with a narrow tabulate venter and sharp ventral shoulders. Flanks slightly convex, gradually converging towards venter. Umbilicus wide, with a steep but very low umbilical wall and rounded umbilical shoulders. Ornamentation composed of straight ribs and strigation on lower flanks only. Ribs arise on umbilical shoulder, and disappear above ventral shoulder. Suture line incompletely known, resembling that of *Flemingites rursiradiatus* but with slightly more phylloid saddles.

Discussion. – Following the definition of Waterhouse (1996b), *Rohillites* is characterized by its narrowly tabulate venter, its extreme lateral compression, its low umbilical wall, and possibly by its more phylloid suture line in comparison to that of *Flemingites*.

R. bruehwileri n. sp. essentially differs from the type species *R. rohilla* by its slightly more involute coiling and a strigation restricted to lower flanks.

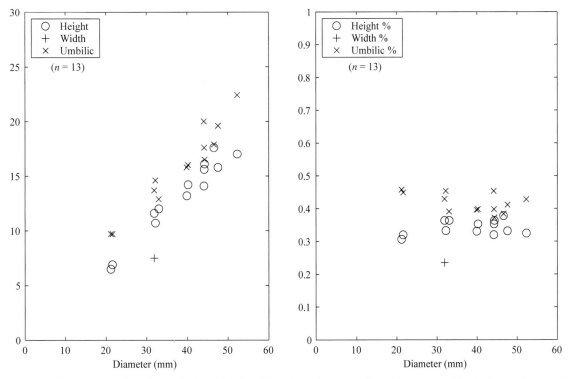

Fig. 41. Scatter diagrams of H, W, and U, and of H/D, W/D and U/D for *Rohillites bruehwileri* n. sp. (Jinya, *Flemingites rursiradiatus* beds).

R. bruehwileri n. sp. also differs from *R. rohilla* by its smaller adult size.

Rohillites sobolevi n. sp.

Pl. 20: 1–2; Fig. 42

Diagnosis. – Rohillites with rursiradiate, low and broad ribs and with markedly converging lower flanks.

Holotype. – PIMUZ 26473, Loc. FSB1/2, Jinya, *Flemingites rursiradiatus* beds, Smithian.

Derivation of name. – Species named after Evgueni S. Sobolev (Institute of Geology, Russian Academy of Science, Novosibirsk).

Occurrence. – Jin4; FSB1/2; *Flemingites rursiradiatus* beds.

Description. – Evolute, laterally compressed shell with a narrow tabulate venter and angular ventral shoulders. Flanks parallel, but becoming markedly convergent on lower flanks. Wide umbilicus with a perpendicular wall and narrowly rounded umbilical shoulders. Ornamentation consists of irregularly spaced and sinuous rursiradiate ribs, becoming very faint on lower flanks and disappearing at ventral

shoulders. The lower flanks exhibit conspicuous, typical strigate ornamentation. Suture incompletely known but resembling that of *R. bruehwileri* n. sp.

Discussion. – R. sobolevi n. sp. is unambiguously distinguished from *R. rohilla* and *R. bruehwileri* n. sp. by its retroverse ribs and plications. This species is distinguished from *R. rohilla* and *R. bruehwileri* by the presence of a blunt angle of the whorl section at two-thirds of flanks, thus clearly demarcating the converging lower flanks. *R. sobolevi* n. sp. and *R. bruehwileri* n. sp. share a similar strigate ornamentation on lower flanks. *R. sobolevi* n. sp. is clearly distinguished from *Flemingites rursiradiatus* by its extreme lateral compression, its narrow flat venter and the angular flank. *R. sobolevi* n. sp. also differs from *F. rursiradiatus* by a strigation restricted to the lower flanks.

Rohillites sp. indet.

Pl. 19: 7a–c

Occurrence. – Jin67; *Kashmirites kapila* beds.

Description. – Evolute, laterally compressed shell with slightly convex flanks, gradually convergent from umbilical shoulder, an extremely thin, tabulate venter, and angular ventral shoulders. Umbilicus

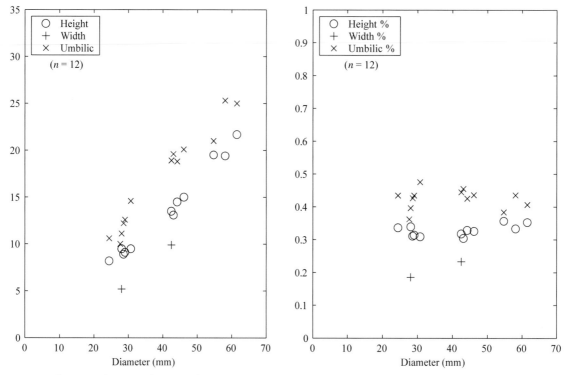

Fig. 42. Scatter diagrams of H, W and U, and of H/D, W/D and U/D for *Rohillites sobolevi* n. sp. (Jinya, *Flemingites rursiradiatus* beds).

fairly wide with moderately high, oblique wall and rounded shoulders. Ornamentation consists of extremely weak plications on inner whorls and sinuous, convex ribs on outer whorls. Sinuous growth lines parallel to the ribs also visible on outer whorls. A portion of flank near venter exhibits conspicuous, typical strigate ornamentation. Suture line unknown.

Discussion. – This single specimen probably represents a new species and can be distinguished from other *Rohillites* by its absence of ornamentation on its inner whorls. However, it is preferable not to erect a new species based on such a very fragmentary specimen.

Genus *Galfettites* n. gen.

Type species. – *Galfettites simplicitatis* n. sp.

Composition of the genus. – Type species only.

Diagnosis. – Laterally compressed Flemingitidae with smooth, flat, parallel flanks and a very narrowly curved venter.

Derivation of name. – Named after Thomas Galfetti (Assistant at PIMUZ).

Occurrence. – *Owenites koeneni* beds.

Discussion. – This genus is clearly distinguished from *Flemingites* and *Euflemingites* by its absence of strigation, and from *Anaxenaspis* and *Pseudoflemingites* by its absence of ribs or plications. Furthermore, *Galfettites* is the only genus within Flemingitidae that exhibits flat, parallel flanks on adult whorls.

Galfettites simplicitatis n. sp.

Pl. 21: 1–2; Fig. 43

Diagnosis. – As for the genus.

Holotype. – PIMUZ 26002, Loc. Jin27, Jinya, *Owenites koeneni* beds, Smithian.

Derivation of name. – From the Latin word: simplicitatis, meaning simplicity in reference to the absence of marked ornamentation.

Occurrence. – Jin27, 45; *Owenites koeneni* beds.

Description. – Very evolute and very compressed shell with a very narrowly curved to subtabulate venter, abruptly rounded ventral shoulders, and flat, parallel flanks for about two-thirds of flank, then

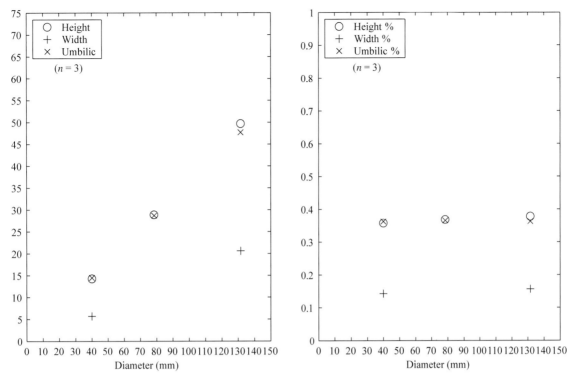

Fig. 43. Scatter diagrams of H, W and U, and of H/D, W/D and U/D for *Galfettites simplicitatis* n. gen., n. sp. (Jinya, *Owenites koeneni* beds).

gradually converging toward venter. Umbilicus wide, with high, nearly perpendicular wall and rounded shoulders. Available specimens do not exhibit any ornamentation. Suture line ceratitic, typical of Flemingitidae, with phylloid saddle only present in umbilical part. Auxiliary series is well developed.

Discussion. – *Galfettites simplicitatis* n. sp. is easily distinguished from other flemingitids by its peculiar whorl section and its absence of ornamentation.

Genus *Pseudoflemingites* Spath, 1930

Type species. – *Ophiceras nopscanum* Welter, 1922

Pseudoflemingites goudemandi n. sp.

Pl. 22: 1–5; Fig. 44

Diagnosis. – Serpenticonic *Pseudoflemingites* with very weak ribbing on juvenile whorls, and a high, ovoid whorl section, without strigation.

Holotype. – PIMUZ 26004, Loc. Jin99, Jinya, *Owenites koeneni* beds, Smithian.

Derivation of name. – Named for Nicolas Goudemand (Assistant at PIMUZ).

Occurrence. – Jin12, 99; NW1; T11; Yu3; *Owenites koeneni* beds.

Description. – Very evolute, compressed serpenticone with a low rounded venter, rounded ventral shoulders, and more or less convex flanks forming a highly variable whorl section ranging from suboval to subrectangular (robust specimens). Umbilicus with moderately high, perpendicular wall and rounded shoulders. Ornamentation consists only of extremely weak, disparate folds on juvenile stages. No strigation observed. Suture line typical for Flemingitidae with three nearly phylloid saddles, broad indented lobes and poorly defined, small auxiliary series.

Discussion. – Although this species does not exhibit ribbing typical of the type species, its mode of coiling, absence of strigation and simple suture line justify its assignment to *Pseudoflemingites*.

Genus *Anaxenaspis* Kiparisova, 1956

Type species. – *Xenaspis orientalis* Diener, 1895

?*Anaxenaspis* sp. indet.

Pl. 23: 1a–b; Table 1

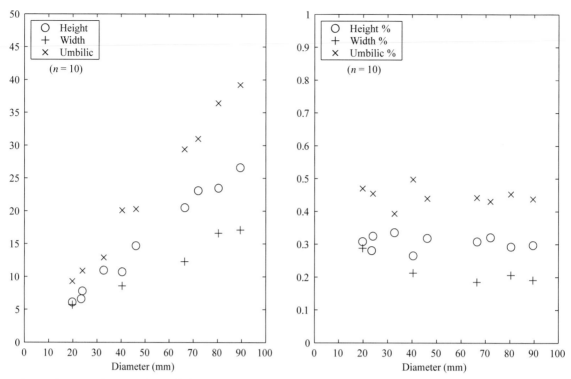

Fig. 44. Scatter diagrams of H, W and U, and of H/D, W/D and U/D for *Pseudoflemingites goudemandi* n. sp. (Jinya, *Owenites koeneni* beds).

Occurrence. – Jin45; *Owenites koeneni* beds.

Description. – Evolute, compressed shell with a sub-tabulate venter, rounded ventral shoulders, convex flanks, and wide umbilicus with moderately high, gently sloping wall and rounded shoulders. Ornamentation consists of dense, forward projected ribs, especially visible on inner whorls. Suture line unknown.

Discussion. – The assignment of this specimen to *Anaxenaspis* is uncertain, and is based only on its similar morphology with Flemingitidae and the absence of strigation.

Genus *Guangxiceras* n. gen.

Type species. – *Guangxiceras inflata* n. sp.

Composition of the genus. – Type species only.

Diagnosis. – Flemingitidae with weak bullae on the inner whorls and very weak folds on the outer whorls. Inner whorls are inflated in contrast to the more compressed outer whorls.

Derivation of name. – Genus name refers to the Guangxi Province (South China).

Occurrence. – *Owenites koeneni* beds.

Discussion. – This genus is clearly distinguished from other Flemingitidae, and especially *Anaxenaspis* and *Pseudoflemingites*, by the weak nodes present on its inner whorls. It also differs by its inflated inner whorls, which are in contrast with its more compressed outer whorls. The suture line agrees in plan with that of Flemingitidae, but displays a compressed umbilical saddle.

Guangxiceras inflata n. sp.

Pl. 23: 2a–e; Table 1

Diagnosis. – As for the genus.

Holotype. – PIMUZ 26016, Loc. Jin27, Jinya, *Owenites koeneni* beds, Smithian.

Derivation of name. – Species name refers to the morphology of the inner whorls.

Occurrence. – Jin27; *Owenites koeneni* beds.

Description. – Laterally compressed shell with nearly parallel flanks on outer whorls and significantly inflated flanks on inner whorls. Inner whorls exhibit serpenticonic coiling, whereas outer whorl is only moderately evolute with respect to next inner whorl. Outer whorl exhibits a subtabulate venter and rounded ventral shoulders. Umbilicus very wide, with moderately high, oblique wall (higher on inner whorls), and rounded shoulders. Ornamentation consists only of weak bullae and folds on inner whorls and very weak folds on outer whorls. Suture line ceratitic with very elongated saddles. Lateral lobe deeply indented and umbilical saddle laterally compressed. Umbilical and lateral saddles slanted slightly towards umbilicus.

Genus *Anaflemingites* Kummel & Steele, 1962

Type species. – *Anaflemingites silberlingi* Kummel & Steele, 1962

Anaflemingites hochulii n. sp.

Pl. 24: 3–6; Fig. 45

Diagnosis. – *Anaflemingites* with highly variable morphology and ornamentation throughout ontogeny. Sinuous growth lines and weak ventrolateral strigation are present on juvenile whorls, while mature specimens exhibit only gentle, fold-type ribs.

Holotype. – PIMUZ 26020, Loc. Jin45, Jinya, *Owenites koeneni* beds, Smithian.

Derivation of name. – Named after Peter A. Hochuli (Professor at PIMUZ).

Occurrence. – Jin42, 45; Yu2; *Owenites koeneni* beds.

Description. – Moderately evolute, laterally compressed shell with a subtabulate venter, abruptly rounded ventral shoulders on juvenile whorls, more gently rounded on larger specimens, and convex flanks exhibiting maximum lateral curvature at mid-flank on juvenile whorls, and more compressed, nearly parallel flanks on mature specimens. Umbilicus wide, with low, nearly perpendicular wall and rounded shoulders. Ornamentation on juvenile whorls consists of variable, sinuous growth lines and weak folds, as well as very weak strigation near venter. Our larger specimens mainly exhibit only conspicuous radial folds. Suture line ceratitic with well-indented lobes and narrow saddles.

Discussion. – Upon comparison of our specimens with the holotype of *A. silberlingi* Kummel & Steele, 1962, we find significant differences that fully justify the erection of a new species. Indeed, strigation

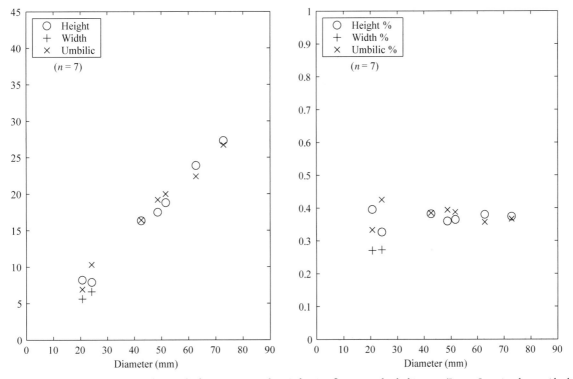

Fig. 45. Scatter diagrams of H, W and U, and of H/D, W/D and U/D for *Anaflemingites hochulii* n. sp. (Jinya, *Owenites koeneni* beds).

appears to be present on the entire flank of the holotype, and not just near the venter as with our specimens. Furthermore, the folds and growth lines exhibited by *A. hochulii* n. sp. are much more conspicuous. The lobes of the suture line illustrated by Kummel & Steele (1962) also appear narrower than for *A. hochulii* n. sp. The distinct morphology of this genus and the presence of strigation have led us to include it within the Flemingitidae, and not the Meekoceratidae as suggested by Tozer (1981) based only on its suture line.

Anaflemingites differs from *Flemingites* and *Rohillites* by its weaker folds and strigation, from *Galfettites* and *Guangxiceras* by the presence of a strigation.

Family Arctoceratidae Arthaber, 1911

Genus *Arctoceras* Hyatt, 1900

Type species: *Ceratites polaris* Mojsisovics, 1886

Arctoceras strigatus n. sp.

Pl. 25: 1–2; Table 1

Diagnosis. – Moderately involute *Arctoceras* without umbilical tuberculation, but with obvious strigation on the entire flank, and two very weak, longitudinal ridges on flank.

Holotype. – PIMUZ 26023, Loc. Jin15, Jinya, *Owenites koeneni* beds, Smithian.

Derivation of name. – Refers to its conspicuous strigation.

Occurrence. – Jin15; FSB1/2; *Owenites koeneni* beds.

Description. – Moderately involute, somewhat thick platycone with a subtabulate to broadly rounded venter, rounded ventral shoulders, and weakly convex flanks convergent to venter. Deep umbilicus with very high, perpendicular wall and abruptly rounded shoulders without tuberculation. Ornamentation consists of very noticeable strigation on flanks, becoming weaker on venter, as well as distant, weak, radial straight folds on lower and mid-flanks. Barely visible on flanks are two, very weak, longitudinal ridges located at about one third and two-thirds of the distance across the flank (Pl. 25: 2c). Suture line ceratitic, typical for Arctoceratidae, with deeply indented lobes and small auxiliary series. Umbilical and lateral saddles slightly slanted towards umbilicus.

Discussion. – This species differs primarily from *A. tuberculatum* (Smith, 1932) by its absence of umbilical tuberculation, but *A. tuberculatum* exhibits a somewhat weaker strigation, and its ribs are more sinuous and even more distant (see Kummel 1961). It is also distinguishable from *A. blomstrandi* (Lindström, 1865) by its less involute coiling, its subtabulate to broadly rounded venter and its more pronounced strigation and ribs. We believe that it may be necessary to revise the definition of *A. blomstrandi* in light of some of the conspicuous differences in ornamentation on specimens illustrated by Kummel (1961). Interestingly, the strigation of *A. strigatus* is more pronounced on the flanks than on the venter.

Arctoceras sp. indet.

Pl. 27: 4a–c

Occurrence. – FW4/5; *Flemingites rursiradiatus* beds.

Description. – Moderately involute, laterally compressed, high whorled platycone with nearly parallel flanks, a subtabulate venter, rounded ventral shoulders, and deep umbilicus with very high, flat, slightly oblique wall and abruptly rounded shoulders. Ornamentation consists of distant, weakly convex ribs, more prominent at mid-flank. No strigation visible. This species is represented only by a fragmentary body chamber. Suture line unknown. Estimated maximum diameter: greater than 15 cm.

Discussion. – This species differs from *A. tuberculatum* (Smith, 1932) by its lack of umbilical tuberculation and strigation, and from *A. strigatus* n. sp. by its absence of strigation. *A. blomstrandi* (Lindström, 1865) is slightly more involute.

Genus *Submeekoceras* Spath, 1934

Type species: *Meekoceras mushbachanum* White, 1879

Submeekoceras mushbachanum (White, 1879)

Pl. 16: 4; Pl. 26: 1–9; Fig. 46

1879	*Meekoceras mushbachanum* White, p. 113.
1880	*Meekoceras mushbachanum* – White, p. 114, pl. 32: 1a–d.
1902	*Prionolobus mushbachanum* – Frech, p. 631: c.
1904	*Meekoceras mushbachanum* – Smith, p. 376, pl. 41: 1–3; pl. 43: 1–2.

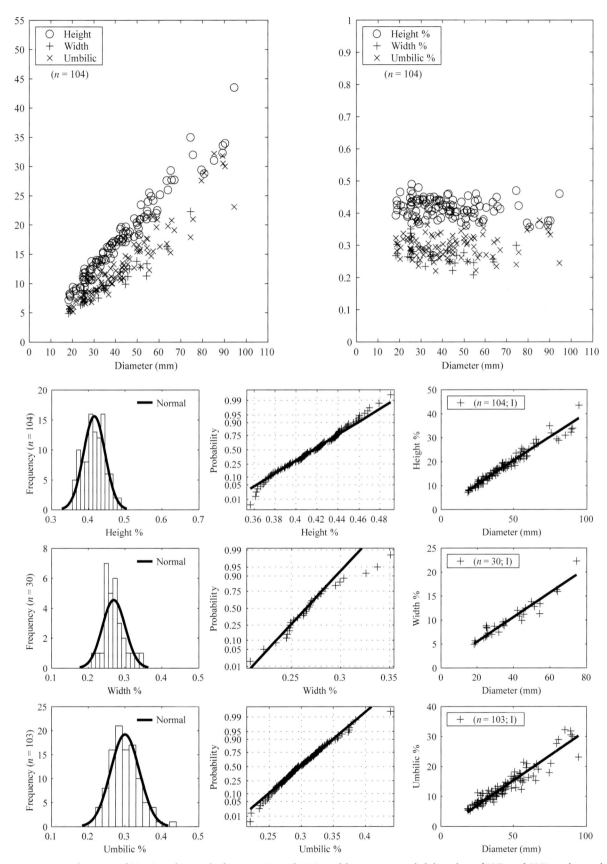

Fig. 46. Scatter diagrams of H, W, and U, and of H/D, W/D and U/D, and histograms, probability plots of H/D and U/D, and growth curves for *Submeekoceras mushbachanum* (Jinya, Waili and Yuping, *Flemingites rursiradiatus* beds). 'A' indicates allometric growth. 'I' indicates isometric growth.

1905   *Meekoceras (Koninckites) mushbachanum*
        – Hyatt & Smith, p. 149, pl. 15: 1–9; pl. 16:
        1–3; pl. 18: 1–7; pl. 70: 8–10.

1914   *Meekoceras mushbachanum* – Smith, p. 77,
        pl. 72: 1–2; pl. 73: 1–6; pl. 74: 1–23.

1915   *Meekoceras mushbachanum* – Diener, p. 193.

non 1922   *Meekoceras mushbachanum* – Welter, p. 126.

1932   *Meekoceras (Koninckites) mushbachanum*
        – Smith, p. 61, pl. 15: 1–9; pl. 16: 1–3;
        pl. 18: 1–7; pl. 38: 1; pl. 59: 17–21; pl. 70:
        8–10; pl. 74: 1–23; pl. 75: 1–6; pl. 76: 1–3.

1932   *Meekoceras (Koninckites) mushbachanum*
        var. *corrugatum* Smith, p. 61, pl. 38: 1.

1932   *Meekoceras (Koninckites) evansi* Smith,
        p. 60, pl. 35: 1–3; pl. 36: 1–18.

1934   *Submeekoceras mushbachanum* – Spath,
        p. 255, fig. 87.

v 1959   *Paranorites ovalis* sp. nov. Chao, p. 217,
        pl. 9: 16–19, text-fig. 16b.

?1959   *Prionolobus ophionus* var. *involutus* var.
         nov. Chao, p. 201, pl. 9: 11–15, text-fig. 11b.

v 1959   *Prionolobus hsüyüchieni* sp. nov. Chao,
          p. 202, pl. 9: 9–10, text-fig. 11c.

?1959   *Meekoceras (Submeekoceras) tientungense*
         sp. nov. Chao, p. 317, pl. 14: 6–7, text-
         fig. 45b.

v 1959   *Meekoceras (Submeekoceras) subquadratum*
          sp. nov. Chao, p. 317, pl. 14: 1–5; pl. 39:
          8–9, text-fig. 45c.

v 1959   *Meekoceras densistriatum* sp. nov. Chao,
          p. 310, pl. 38: 1–3, 19, text-fig. 43b.

v 1959   *Meekoceras yukiangense* sp. nov. Chao,
          p. 311, pl. 39: 1–7, text-fig. 44a.

v 1959   *Meekoceras kaohwaiense* sp. nov. Chao,
          p. 311, pl. 40: 16–18, text-fig. 44b.

v 1959   *Meekoceras pulchriforme* sp. nov. Chao,
          p. 313, pl. 40: 14–15, text-fig. 44c.

?1959   *Meekoceras jolinkense* Chao, p. 314, pl. 14:
         12–15.

v 1959   *Meekoceras lativentrosum* sp. nov. Chao,
          p. 309, pl. 38: 15–18, text-fig. 43a.

v 1959   *Proptychites latumbilicutus* sp. nov. Chao,
          p. 234, pl. 19: 2–3, text-fig. 22a.

?1959   *Proptychites kaoyunlingensis* sp. nov. Chao,
         p. 234, pl. 16: 7–8, text-fig. 22b.

p 1968   *Arctoceras mushbachanum* – Kummel &
          Erben, p. 131 (*pars*), pl. 21: 1 only.

Occurrence. – Jin4, 11, 13, 22, 23, 24, 26, 28, 29, 30,
41, 51; FSB1/2; WFB; FW4/5; Sha1; T6, T50; *Flemingites rursiradiatus* beds. Jin10, 27; *Owenites koeneni*
beds.

Description. – Slightly evolute, somewhat compressed platycone with flat, parallel flanks, a broadly
rounded to subtabulate venter and rounded ventral
shoulders. Deep umbilicus with very high, flat,
perpendicular wall and abruptly rounded shoulders.
Ornamentation consists of sinuous growth lines on
all specimens, and weak, but noticeable fold-type ribs
similar to *Arctoceras*, on a few specimens.

The morphology of juvenile specimens is similar
to *Arctoceras tuberculatum*, but without umbilical
tubercles. Mature specimens are more evolute and
somewhat more laterally compressed. Suture line
ceratitic with three principal saddles and an auxiliary
series. As already noticed by Kummel & Erben (1968:
p. 133), an examination of numerous specimens
reveals considerable variation in the suture lines.
This variation is also true for the genus *Arctoceras*.
However, the lateral lobe of *Arctoceras* is always
more deeply indented. The auxiliary series and
saddles become longer with ontogenetic growth.

Discussion. – This species clearly belongs to the
Arctoceratidae. However, with the exception of a few
species such as *A. tuberculatum*, differentiation
between species can be difficult for juvenile members
of the family. The specimen illustrated by Kummel &
Erben (1968, pl. 21: 2) is too involute to be referred
to as *S. mushbachanum*. Whorl height, width and
umbilical diameter of *S. mushbachanum* exhibit
isometric growth (Fig. 46). Estimated largest diameter
exceeding 15 cm.

Family Ussuridae Spath, 1930

Genus *Ussuria* Diener, 1895

Type species. – *Ussuria schamarae* Diener, 1895

Ussuria kwangsiana Chao, 1959

Pl. 27: 1–3; Fig. 47

1959   *Ussuria kwangsiana* Chao, p. 258, pl. 31: 8–
        10, text-fig. 30a.

?1959   *Ussuria pakungiana* sp. nov. Chao, p. 258,
         pl. 31: 1–3, text-fig. 30c, d.

?1959   *Ussuria longilobata* sp. nov. Chao, p. 259,
         pl. 31: 4–7, text-fig. 30b.

Occurrence. – Jin27, 45; *Owenites koeneni* beds.

Description. – Extremely involute, compressed oxycone
with a very narrowly curved venter, becoming somewhat acute on larger specimens, a rapidly expanding
whorl height, and convex flanks with maximum

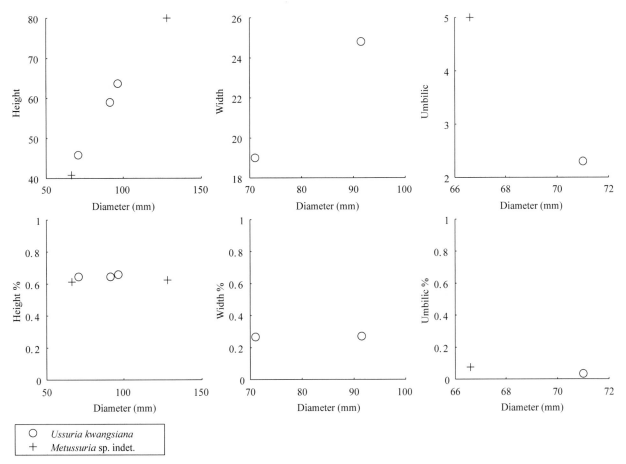

Fig. 47. Scatter diagrams of H, W, and U, and of H/D, W/D and U/D for *Ussuria kwangsiana* and *Metussuria* sp. indet. (Jinya, *Owenites koeneni* beds).

thickness near the umbilicus. Umbilicus nearly occluded, but deep with high, almost perpendicular wall and abruptly rounded shoulders. Ornamentation consists only of very weak plications. Very fine radial growth lines are displayed. Suture line subammonitic, typical of genus. Ventral lobe deeply indented, all saddles are monophylloid and narrower than lobes. Simple auxiliary series not completely preserved, but is visible.

Discussion. – This species is more laterally compressed than the type species *U. schamarae. Ussuria* superficially resembles *Pseudosageceras*. However, the bicarinate venter and the suture line of *Pseudosageceras* are highly distinctive.

Genus *Metussuria* Spath, 1934

Type species. – *Ussuria waageni* Hyatt & Smith, 1905

Metussuria sp. indet.

Pl. 25: 3–8; Fig. 47

Occurrence. – Jin27; *Owenites koeneni* beds.

Description. – Very involute, very compressed oxycone with a narrowly rounded venter (more or less subtabulate for juveniles) and convex flanks with maximum lateral curvature near umbilicus, gradually convergent to venter. Umbilicus nearly occluded with low, oblique wall and rounded shoulders. Ornamentation consists only of weak folds. Thin, radial growth lines are visible. Suture line with almost diphylloid saddles and deeply indented lobes, adventious elements and a complex auxiliary series. Estimated largest diameter exceeding 20 cm.

Discussion. – The morphology and measurements of *Metussuria* sp. indet. are very close to those of *U. kwangsiana*, but it can be distinguished by its more complex and deeply indented suture line, its slightly more compressed whorl section, and its somewhat shallower umbilicus. *Metussuria* sp. indet. is more compressed than the type species, and it is difficult to definitely assign it to *M. spathi* Chao, 1959, since the illustrations given by Chao are insufficient.

Metussuria essentially differs from *Parussuria* Spath, 1934 by the absence of strigation.

Genus *Parussuria* Spath, 1934

Type species. – *Ussuria compressa* Hyatt & Smith, 1905

Parussuria compressa (Hyatt & Smith, 1905)

Pl. 12: 17

  1905   *Ussuria compressa* sp. nov. Hyatt & Smith, p. 89, pl. 3: 6–11.
  1932   *Sturia compressa* – Smith, p. 93, pl. 3: 6–11.
  1934   *Parussuria compressa* – Spath, p. 213, fig. 66c, d.
  1962   *Parussuria compressa* – Kummel & Steele, p. 690, pl. 99: 23; pl. 102: 11.
?1968   *Parussuria semenovi* sp. nov. Zakharov, p. 59, pl.5: 4.
  1995   *Parussuria compressa* – Shevyrev, p. 37, pl. 4: 6, text-fig. 16.

Occurrence. – Jin27; *Owenites koeneni* beds.

Description. – Very involute, compressed oxycone with a narrowly rounded venter and slightly convex flanks with maximum lateral curvature near umbilicus, gradually converging to venter. Umbilicus nearly occluded, with moderately high, oblique wall and rounded shoulders. Ornamentation consists of weak strigation. Thin, radial growth lines are visible. Suture line unknown due to fragmentary preservation of only specimen.

Discussion. – This genus is very similar to *Ussuria* and *Metussuria*, but differs by its more indented, complex suture line, and by its weak strigation. *P. latilobata*, assigned to the Ussuriidae by Chao (1959), is questionable because it was found in the Spathian, which supposedly does not include Ussuriidae. Furthermore, the suture line of *P. latilobata* appears to be either poorly preserved, or it has been excessively ground away during preparation. All measurements of *Ussuria*, *Metussuria* and *Parussuria* are quite close, thus indicating the probability of very strong phylogenetic affinities. This similarity also suggests that it can be very difficult to distinguish between species of the family, unless the suture line and/or subtle ornamentation (e.g. strigation) is preserved.

Family Prionitidae Hyatt, 1900

Genus *Anasibirites* Mojsisovics, 1896

Type species. – *Sibirites kingianus* Waagen, 1895

Anasibirites multiformis Welter, 1922

Pl. 28: 1–6; Figs. 48, 49

p ?1895   *Sibirites tenuistriatus* n. sp. Waagen, p. 124, pl. 9: 2a–b.
p 1922   *Anasibirites multiformis* nov. sp. Welter, p. 138, pl. 15: 12-13, 23–24; pl. 16: 6–19.
   ?1929   *Anasibirites welleri* n. sp. Mathews, p. 14, pl. 2: 17–19.
   ?1929   *Anasibirites emmonsi* n. sp. Mathews, p. 14, pl. 2: 20–26.

Occurrence. – Jin16, 48, 100, 101; FW6; NW12; *Anasibirites multiformis* beds.

Description. – Moderately evolute, compressed platy-cone with a subtabulate to tabulate venter, angular to subangular ventral shoulders, and slightly convex flanks with maximum thickness near umbilicus for juveniles and mid-flank for mature specimens. Moderately deep umbilicus with oblique wall and rounded shoulders. Upon comparison of various species of *Anasibirites*, it becomes readily apparent that it must be redefined according to its characteristic, but highly variable ornamentation (see discussion):

• very few distinct ribs
• dense, concave and thick striae: exhibited by all developmental stages, strongly forward projected, crossing venter without significant deviation

In contrast with other *Anasibirites* species, its ribs are not pronounced and they do not alternate with weak ribs and growth lines. Ceratitic suture line is typical of prionitids.

Discussion. – It is extremely difficult to distinguish various species of *Anasibirites* from each other due to their similarity in suture lines and extreme variability of ornamentation, as well as a lack of measurements for previously illustrated species. Quite often, different workers (e.g. Kummel & Erben 1968) concluded that all previously described species should be placed within the synonymy of a single species related to the type species: *A. kingianus*. However, a comprehensive study of the various types of ornamentation and umbilical characteristics suggests the existence of at least four main species:

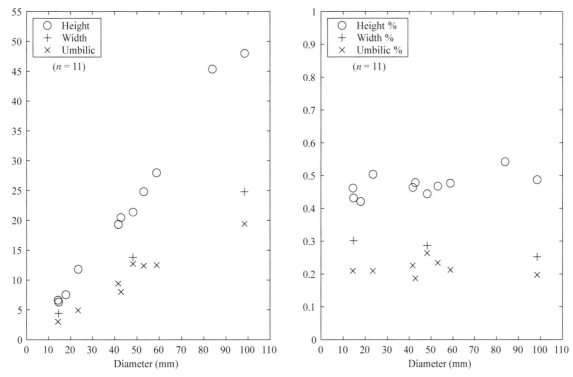

Fig. 48. Scatter diagrams of H, W and U, and of H/D, W/D and U/D for *Anasibirites multiformis* (Jinya and Waili, *Anasibirites multiformis* beds).

1 *A. kingianus* (Waagen, 1895): Distinguishing characteristics include an obviously arched venter and ornamentation consisting of an alternation of sinuous, weak and strong, forward projecting ribs. Ribs strongly attenuated on adult specimens.
2 *A. pluriformis* Guex, 1978: with a tabulate to subtabulate venter, a perpendicular umbilical wall, and ornamentation consisting of weak and/ or distinct ribs forming umbilical and ventral tuberculations on some robust specimens. Ribs are more radial.
3 *A. multiformis* Welter, 1922 (redefined here): with a tabulate to subtabulate venter and identical ornamentation on all developmental stages (dense, concave, forward projected growth lines with very few distinct ribs).
4 *A. nevolini* Burij & Zharnikova, 1968: It exhibits forward projected growth lines, and a regular alternation of concave, weak and strong ribs, as well as more evolute coiling at small sizes. In contrast to *A. pluriformis*, robust variants do not exhibit umbilical tubercules. This alternation of ribbing strength is somewhat attenuated on adult specimens, thus making it even more difficult to distinguish it from other interpretation.

Unfortunately, the lack of measurements for many previously described species precludes a definitive validation of this new interpretation.

Discussion. – A. multiformis displays a very strong resemblance to *Sibirites tenuistriatus* (Waagen, 1895), and may, in fact, be a synonym of this species. The various species attributed to *Anasibirites* by Chao (1959) do not correspond to this genus, with the exception of *A. multiformis* var. *alternatus*. *A. multiformis* represents a simplified theme of ornamentation within the genus *Anasibirites*.

Anasibirites nevolini Burij & Zharnikova, 1968

Pl. 28: 7–9; Fig. 49

p ?1922 *Anasibirites multiformis* nov. sp. Welter, p. 138, pl. 15: 9–11, 14–16, 19–20, 25–27.
 ?1929 *Anasibirites alternatus* n. sp. Mathews, p. 23, pl. 4: 22–23.
 ?1929 *Anasibirites romeri* n. sp. Mathews, p. 23, pl. 4: 24–25.
 ?1929 *Anasibirites gibsoni* n. sp. Mathews, p. 29, pl. 5: 4–5.
 ?1959 *Anasibirites multiformis* var. *alternatus* Matthews Chao, p. 328, pl. 40: 11.
 ?1964 *Anasibirites pacificus* Bando, p. 73, pl. 3: 5–7; pl. 5: 8, 11, 13, 14; pl. 6: 8, 9, 11.
 ?1964 *Anasibirites ehimensis* n. sp. Bando, p. 74, pl. 3: 12a–c.
p 1968 *Anasibirites nevolini* Burij & Zharnikova *in* Zakharov, p. 131, pl. 25: 5.

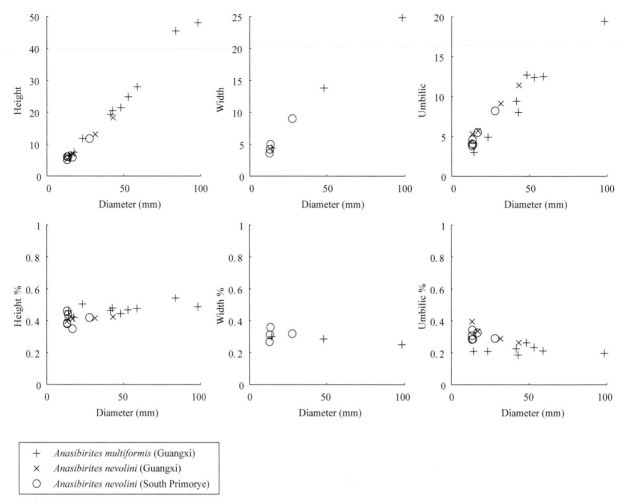

Fig. 49. Scatter diagrams of H, W, and U, and of H/D, W/D and U/D for comparison of *Anasibirites nevolini* and *A. multiformis* (Jinya and Waili, *Anasibirites multiformis* beds; South Primorye data from Zakharov 1968).

1978 *Anasibirites nevolini* – Zakharov, pl. 11: 9–13.

Occurrence. – Jin16, 48; FW6; *Anasibirites multiformis* beds.

Description. – Moderately evolute, compressed platycone with a tabulate to subtabulate venter, subangular to abruptly rounded ventral shoulders, and slightly convex flanks with maximum curvature near umbilicus. Umbilicus with moderately high, oblique wall and rounded shoulders. Ornamentation consisting of a regular alternation of weak and strong concave projected ribs. Stronger ribs do not form umbilical and ventral tubercules. Ornamentation somewhat attenuated on adult specimens. Suture line identical to that of other species of *Anasibirites*.

Discussion. – Juvenile stages are more evolute in *A. nevolini* than in *A. multiformis*. *A. nevolini* also displays a regular alternation of concave thick and

thin ribs. Although this species is relatively rare, it is distinctively more evolute than all other congeneric species (see Fig. 49).

Genus *Hemiprionites* Spath, 1929

Type species. – *Goniodiscus typus* Waagen, 1895

Hemiprionites cf. butleri (Mathews, 1929)

Pl. 29: 1–7; Fig. 50

1929 *Goniodiscus butleri* n. sp. Mathews, p. 35, pl. 6: 18–21.

Occurrence. – Jin16, 48; FW6; *Anasibirites multiformis* beds.

Description. – Involute, compressed shell with a tabulate venter on juvenile whorls, a tabulate or bicarinate venter on mature specimens, angular

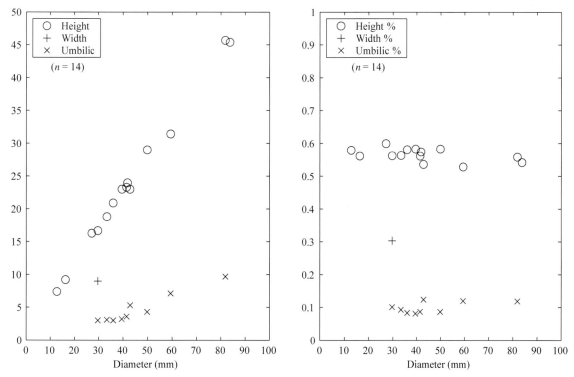

Fig. 50. Scatter diagrams of H, W and U, and of H/D, W/D and U/D for *Hemiprionites* cf. *butleri* (Jinya and Waili, *Anasibirites multiformis* beds).

ventral shoulders, and convex flanks with maximum lateral curvature at mid-flank. Umbilicus extremely narrow at juvenile stages, becoming more open, with noticeable egression at adult stages. On adult specimens, umbilicus has high, gently inclined wall and rounded shoulders. Ornamentation on inner whorls similar to *Anasibirites* and exhibits growth lines crossing the venter. On adult stages, growth lines are more visible, and some develop into weak folds. Suture line ceratitic, similar to *H. typus* with three broad saddles and a small auxiliary series.

Discussion. – *H. butleri* differs from the type species of the genus by its more involute coiling, its greater whorl height and its bicarinate venter on some specimens. Its coiling is significantly more involute than *Anasibirites*. Other species described by Mathews in 1929, with the exception of *H. walcotti*, can be grouped together as variants of the type species. This genus, as is true for all prionitids, displays a great morphological variation. Chinese specimens are morphologically close to *H. butleri*, but it is not possible to firmly assign them to this species, given the lack of a sufficient number of measured American specimens.

Hemiprionites klugi n. sp.

Pl. 30: 1–4; Fig. 51

Diagnosis. – Extremely involute *Hemiprionites* throughout ontogeny, displaying an ovoid whorl section and egressive coiling on adult stages.

Holotype. – PIMUZ 26062, Loc. FW6, Waili, *Anasibirites multiformis* beds, Smithian.

Derivation of name. – Named after Christian Klug (Assistant Professor at PIMUZ).

Occurrence. – Jin16; FW6; *Anasibirites multiformis* beds.

Description. – Very involute throughout ontogeny, with slight egressive coiling on mature specimens. Compressed shell with an ovoid whorl section, a tabulate venter (thinner than *H.* cf. *butleri*), rounded ventral and umbilical shoulders, and convex flanks with maximum curvature at mid-flank. Umbilicus with high, perpendicular wall. No visible ornamentation. Suture line differs from type species, but retains same number of elements. Auxiliary series shorter than for type species.

Discussion. – *H. klugi* n. sp. differs from *H.* cf. *butleri* by its ovoid whorl section, its thiner venter and its peculiar suture line. The suture line of *H. klugi* n. sp. may be similar to some species of *Wasatchites*. *H. walcotti* Mathews, 1929 is similar to our Chinese

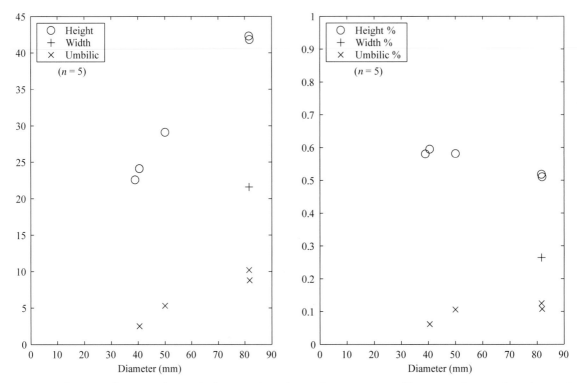

Fig. 51. Scatter diagrams of H, W and U, and of H/D, W/D and U/D for *Hemiprionites klugi* n. sp. (Jinya and Waili, *Anasibirites multiformis* beds).

specimens, but the lack of measurements prevents comparison with *H. klugi* n. sp. Furthermore, the suture line of *H. walcotti* is more similar to the type species than to *H. klugi* n. sp.

Family Inyoitidae Spath, 1934

Genus *Inyoites* Hyatt & Smith, 1905

Type species. – *Inyoites oweni* Hyatt & Smith, 1905

Inyoites krystyni n. sp.

Pl. 31: 1–4; Pl. 32: 1–2; Fig. 52

?1959 *Subvishnuites tientungensis* sp. nov. Chao, p. 210, pl. 7: 17–18.
?1959 *Subvishnuites* sp. undet. Chao, p. 210, pl. 44: 9–10.

Occurrence. – Jin12, 99; Yu3; *Owenites koeneni* beds.

Diagnosis. – Large-sized and very evolute *Inyoites* with a conspicuous, lanceolate venter and very weak, rursiradiate folds.

Holotype. – PIMUZ 26067, Loc. Yu3, Yuping, *Owenites koeneni* beds, Smithian.

Derivation of name. – Named after Leopold Krystyn (University of Vienna).

Description. – Very evolute, compressed shell with a lanceolate venter and a conspicuous keel (present only when outer shell is preserved), rounded ventral shoulders, and parallel flanks. Umbilicus very wide, with moderately high, oblique wall and rounded shoulders. Ornamentation composed of dense, rursiradiate folds on inner whorls, becoming weaker on mature specimens. Suture line ceratitic with indented lobes and three large saddles on adult specimens.

Discussion. – *I. krystyni* n. sp. can be essentially distinguished from other *Inyoites* species by its large size, its more evolute coiling, and its weaker ornamentation.

Genus *Subvishnuites* Spath, 1930

Type species. – *Vishnuites* spec. Welter, 1922 (= *Subvishnuites welteri* sp. nov. Spath, 1930).

Subvishnuites stokesi (Kummel & Steele, 1962)

Pl. 29: 8a–d; Table 1

1962 *Inyoites stokesi* n. sp. Kummel & Steele, p. 672, pl. 99: 19–22.

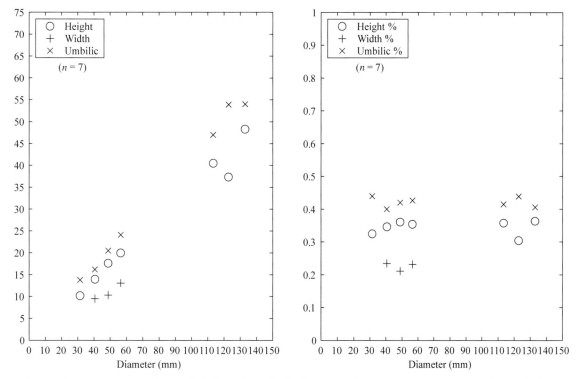

Fig. 52. Scatter diagrams of H, W and U, and of H/D, W/D and U/D for *Inyoites krystyni* n. sp. (Jinya, *Owenites koeneni* beds).

?1973   *Inyoites stokesi* – Collignon, p. 137, pl. 1: 10, 10a.

Occurrence. – Jin12; *Owenites koeneni* beds.

Description. – Compressed serpenticone with a very angular venter, a wide, fairly shallow umbilicus with rounded shoulders, and slightly concave flanks converging towards venter. Thin, rursiradiate folds most prominent on the dorsal half of whorl, but disappear towards venter. Suture line saddles rounded and indented with an auxiliary series composed of small denticulations.

Discussion. – *S. stokesi* is also found in the *Meekoceras* beds of California, Nevada, Utah and Idaho. This species was first assigned to *Inyoites*, but it differs from *I. oweni* Hyatt & Smith, 1905 by its lack of a hollow keel and its angular venter. Since its general shape is closer to the genus *Subvishnuites* Spath, 1930, it is preferable to assign this species to *Subvishnuites*.

Subvishnuites was tentatively placed within the Xenoceltidae by Tozer (1981). However, it has much more in common with the Inyoitidae (e.g. its angular venter), and is not compatible with the morphologically simple Xenoceltidae.

Family Lanceolitidae Spath, 1934

Genus *Lanceolites* Hyatt & Smith, 1905

Type species. – *Lanceolites compactus* Hyatt & Smith, 1905

Lanceolites compactus Hyatt & Smith, 1905

Pl. 30: 5a–c; Fig. 53

  1905   *Lanceolites compactus* sp. nov. Hyatt & Smith, p. 113, pl. 4: 4–10; pl. 5: 7–9; pl. 78: 9–11.
  1932   *Lanceolites compactus* – Smith, p. 90, pl. 4: 4–10; pl. 5: 7–9; pl. 21: 21–23; pl. 28: 17–20; pl. 40: 9–11; pl. 60: 10.
  1962   *Lanceolites compactus* – Kummel & Steele, p. 692, pl. 102: 6–9.
?1979   *Lanceolites compactus* – Nichols & Silberling, pl. 2: 39–43.
  1995   *Lanceolites compactus* – Shevyrev, p. 39, pl. 2: 1–2.

Occurrence. – Jin12; T5; *Owenites koeneni* beds.

Description. – Extremely involute, discoidal shell with a narrow, concave, bicarinate venter and slightly

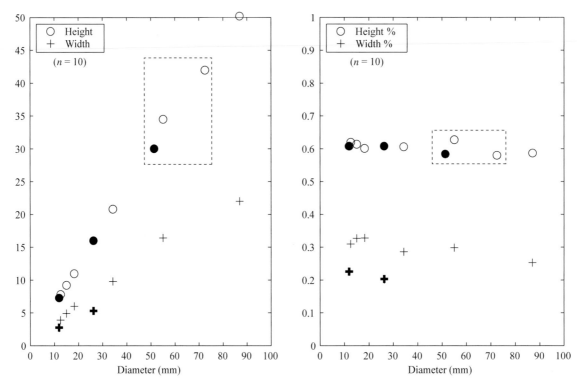

Fig. 53. Scatter diagrams of H, W, and U, and of H/D, W/D and U/D for *Lanceolites bicarinatus* and *L. compactus*. Solid circles indicate *L. bicarinatus*; open circles indicate *L. compactus*. Area bounded by dotted line represents specimens from Guangxi, Jinya, *Owenites koeneni* beds. Other data from Vu Khuc 1984 and Shevyrev 1995.

convex flanks, more convergent on outer half of whorl. Umbilicus is occluded and shell exhibits a rapidly expanding whorl width. No visible ornamentation on our specimens. Suture line, although poorly preserved on our specimens, is typical of that of the genus with its single, broad and indented lateral lobe.

Discussion. – This species can be distinguished from *L. bicarinatus* by its greater whorl width.

Lanceolites bicarinatus Smith, 1932

Pl. 30: 6a–d; Fig. 53

    1932 *Lanceolites bicarinatus* Smith, p. 90, pl. 55: 1–13.
v 1959 *Lanceolites orientalis* sp. nov. Chao, p. 263, pl. 41: 5–9.
    1984 *Lanceolites bicarinatus* – Vu Khuc, p. 85, pl. 7: 2a–b, text-fig. H18.
    1995 *Lanceolites bicarinatus* – Shevyrev, p. 40, pl. 4: 3.

Occurrence. – Jin12, 45; Yu7; *Owenites koeneni* beds.

Description. – Extremely involute, very compressed, discoidal shell with a lenticular whorl section, a

concave, bicarinate venter, angular ventral shoulders, and convex flanks with maximum lateral curvature one third of distance across flank from umbilicus. Umbilicus occluded, with rapidly expanding whorl width, totally embracing penultimate volution. No ornamentation observed. Suture line very peculiar, but typical of genus with very broad, deeply indented lobe followed by several auxiliary elements. All saddles are narrow.

Discussion. – The morphology of the specimen illustrated by Chao (1959) is very similar, but its suture line is quite different. This may be the result of excessive grinding during laboratory preparation. *L. bicarinatus* essentially differs from the type species by its distinctive, bicarinate venter.

Family Paranannitidae Spath, 1934

Genus *Paranannites* Hyatt & Smith, 1905

Type species. – *Paranannites aspenensis* Hyatt & Smith, 1905

Paranannites aff. *aspenensis* Hyatt & Smith, 1905

Pl. 33: 1–10; Fig. 54

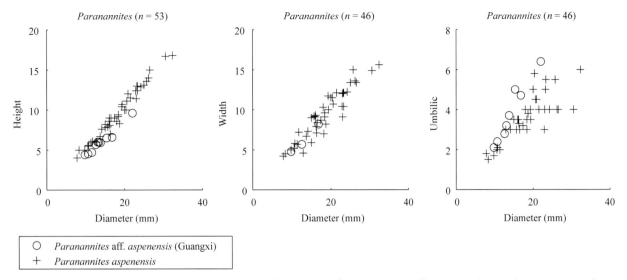

Fig. 54. Scatter diagrams of H, W and U against corresponding diameter for *Paranannites* aff. *aspenensis* (Jinya, *Flemingites rursiradiatus* beds; *P. aspenensis* is given for comparison: data from Kummel & Steele 1962).

v 1959　*Paranannites* cf. *aspenensis* – Chao, p. 284, pl. 24: 1–7, 11, 12.

v ?1959　*Paranannites ptychoides* sp. nov. Chao, p. 284, pl. 24: 8–10, text-fig. 37a.

?1966　*Paranannites aspenensis* – Hada, p. 112, pl. 4: 5a–b.

Occurrence. – Jin4, 13, 23, 28, 29, 30; FW2, 3, 4, 5; *Flemingites rursiradiatus* beds. Jin10; *Owenites koeneni* beds.

Description. – Globose, involute, slightly compressed paranannitid with an arched venter, rounded ventral and umbilical shoulders, and parallel flanks near umbilicus, then gradually convergent to venter. Umbilicus with high, perpendicular wall. Ornamentation extremely variable with weak, prorsiradiate constrictions and folds. A few thin growth lines can be observed. Suture line ceratitic with two broad saddles.

Discussion. – The measurements of our specimens are close to *P. aspenensis* (Fig. 54), and the suture line is also similar, but it is not possible to definitely assign them to this species. Indeed, they differ somewhat from *P. aspenensis* by having a wider umbilicus and they are slightly more depressed. They differ in much the same way from *P. hindostanus* (Diener, 1897, pl. 7: 3a–b). Our specimens are also similar to *P. mulleri* (Kummel & Steele, 1962), but the latter's coiling is more egressive. *P. ptychoides* Chao, 1959 cannot be separated from *P. aspenensis* as proposed by Kummel & Erben (1968), based only on the small denticulations of the suture line on the umbilical wall.

Paranannites spathi (Frebold, 1930)

Pl. 35: 10–19; Fig. 55

1930　*Prosphingites spathi* Frebold, p. 20, pl. 4: 2–3, 3a.

1934　*Prosphingites spathi* – Spath, p. 195, pl. 13: 1a–e, 2.

p ?1959　*Prosphingites kwangsianus* sp. nov. Chao, p. 296, pl. 28: 17–18.

p ?1959　*Prosphingites sinensis* sp. nov. Chao, p. 297, pl. 27: 14–17, text-fig. 40a.

?1982　*Prosphingites spathi* – Korchinskaya, pl. 5: 2a–b

?1994　*Paranannites spathi* – Tozer, p. 77, pl. 36: 1–2.

Occurrence. – Jin27, 43, 45; Yu1; *Owenites koeneni* beds.

Description. – Moderately evolute, subglobular shell with a subangular to rounded (as for some juveniles) venter and convex flanks, gradually converging to venter from abruptly rounded umbilical shoulder. Deep, crateriform umbilicus with high, perpendicular wall, wall height increasing proportionally with diameter. Ornamentation similar, but typical of *Paranannites*-type species with constrictions of variable strength. Constrictions may possibly correspond to megastriae as indicated by presence of a marked ventral sinus on one juvenile specimen. Suture line ceratitic, similar to *P. aspenensis*, with two main, broad saddles. Lobes finely indented and an isolated auxiliary saddle is present.

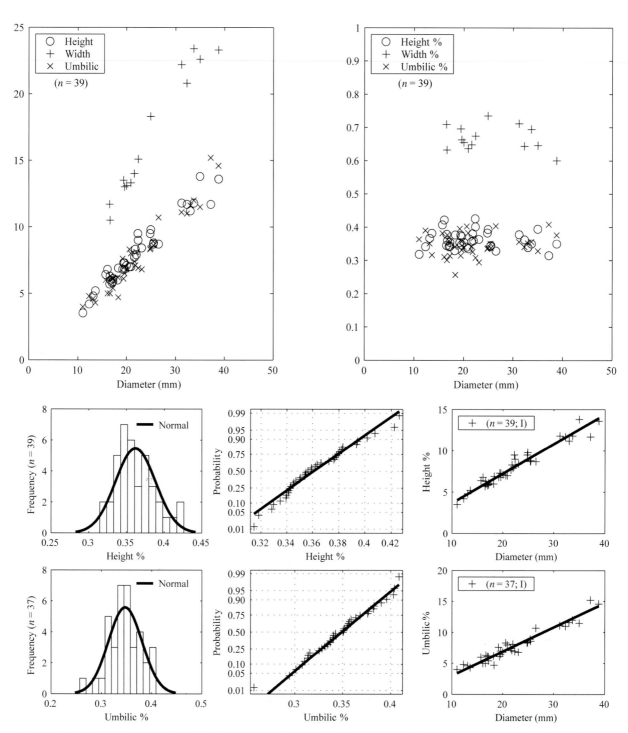

Fig. 55. Scatter diagrams of H, W and U, and of H/D, W/D and U/D, and histograms, probability plots of H/D, U/D and growth curves for *Paranannites spathi* (Jinya and Yuping, *Owenites koeneni* beds). 'A' indicates allometric growth. 'I' indicates isometric growth.

Discussion. – P. spathi differs from other *Paranannites* species by its subangular venter and deep, crateriform umbilicus. Whorl height and umbilical diameter exhibit isometric growth (Fig. 55). *P. slossi* (Kummel & Steele, 1962) exhibits a greater whorl height and narrower width. Some of the specimens described by Chao (1959) as *Prosphingites kwangsianus*

and *Prosphingites sinensis* may actually correspond to *P. spathi*. Tozer (1994) provided the most recent description of *P. spathi*. However, these appear to be too rounded and laterally compressed to be assigned to this species. The same observation can be made regarding specimens described by Korchinskaya (1982) from Spitsbergen. However, Korchinskaya's

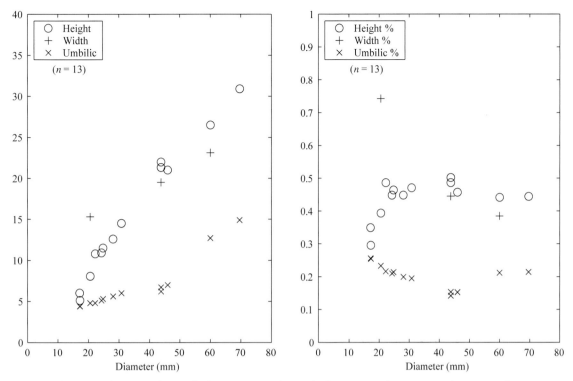

Fig. 56. Scatter diagrams of H, W and U, and of H/D, W/D and U/D for '*Paranannites*' *ovum* n. sp. (Tsoteng and Yuping, *Owenites koeneni* beds).

and Tozer's specimens could also be extreme, laterally compressed variants of *P. spathi*.

'*Paranannites*' *ovum* n. sp.

Pl. 34: 1–6, Fig. 56

Diagnosis. – Large-sized *Paranannites* with globular inner whorls having a broadly arched venter, a compressed high-whorled shape at maturity and a suture line similar to genus *Owenites*.

Holotype. – PIMUZ 26087, Loc. Yu1, Yuping, *Owenites koeneni* beds, Smithian.

Derivation of name. – Refers to its ovoid mature shape.

Occurrence. – Jin45; Yu1; T8; *Owenites koeneni* beds.

Description. – Inner whorls involute and globular with broadly arched venter. Larger specimens less involute, significantly more compressed with rounded venter and gently convergent flanks from umbilical shoulder, becoming more convergent near venter. Umbilicus small, but deep with perpendicular wall and abruptly rounded shoulders. Coiling tends to be

egressive at maturity. Ornamentation consists of very fine, forward projected plications that fade on venter, as well as very fine, growth lines visible only on largest specimens. Suture line ceratitic, similar to genus *Owenites*, with four saddles, and well-crenulated lobes. Ventral saddle with small indentation on sides.

Discussion. – *P. ovum* n. sp. can be distinguished from other *Paranannites* by its larger adult size and its more compressed shell shape. Its suture line also exhibits more elements than the type species of *Paranannites* and is close to *Owenites*.

Paranannites subangulosus n. sp.

Pl. 35: 1–9; Fig. 57

p ?1959 *Paranannites involutus* sp. nov. Chao, p. 297, pl. 24: 13–14.

Diagnosis. – Very involute *Paranannites* with a globular shape and a subangular venter on some specimens.

Holotype. – PIMUZ 26094, Loc. Jin30, Jinya, *Flemingites rursiradiatus* beds, Smithian.

Derivation of name. – Named for its subangular venter.

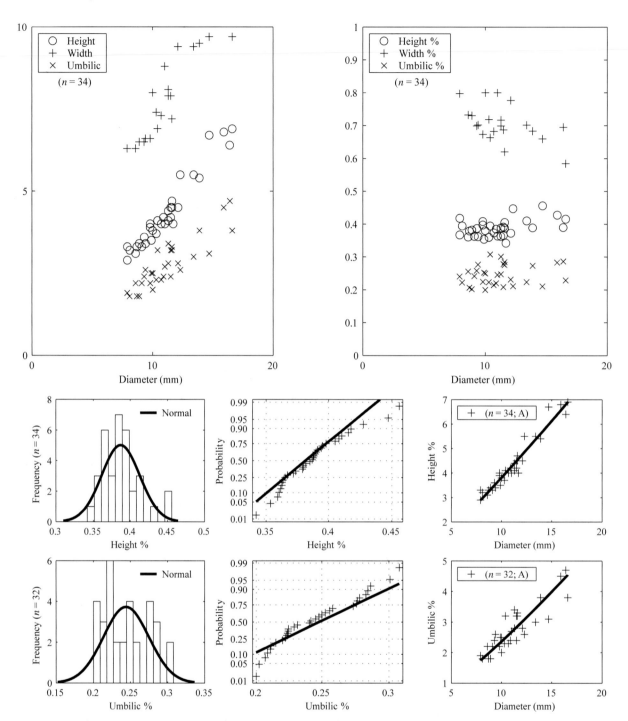

Fig. 57. Scatter diagrams of H, W and U, and of H/D, W/D and U/D, and histograms, probability plots of H/D, U/D and growth curves for *Paranannites subangulosus* n. sp. (Jinya, *Flemingites rursiradiatus* beds). 'A' indicates allometric growth. 'I' indicates isometric growth.

Occurrence. – Jin4, 23, 24, 28, 29, 30, 41; FSB1/2; Sha1; T6, T50; *Flemingites rursiradiatus* beds.

Description. – Small, unusual, very involute *Paranannites* with a globular shape. Venter rounded on most specimens, but a few, more compressed variants may exhibit a subangular venter. Umbilicus deep with perpendicular wall and rounded shoulders.

Body chamber length greater than one whorl. No visible ornamentation. Ceratitic suture line with wide saddles, very typical of Paranannitidae.

Discussion. – This species represents an unusual morphology among the Paranannitidae, as demonstrated by its two extreme variants in venter shape (circular to subangular), but it clearly belongs to this family as evidenced by its suture line. Measurements of *P. subangulosus* n. sp. indicate that this species exhibits strong allometric growth, especially for whorl width (Fig. 57). A rapid increase in width is readily apparent on medium-sized specimens. Measurements also indicate a large intraspecific variation. This species is closely linked by its morphology and suture line to the genus *Thermalites*. However, its suture line is apparently more divided.

Paranannites dubius n. sp.

Pl. 33: 11–14; Table 1

Diagnosis. – Extremely involute *Paranannites* with conspicuous egressive coiling on mature specimens.

Holotype. – PIMUZ 26084, Loc. Jin4, Jinya, *Flemingites rursiradiatus* beds, Smithian.

Derivation of name. – From the Latin word: dubius, meaning doubtful in the sense of not conforming to a pattern.

Occurrence. – Jin4; *Flemingites rursiradiatus* beds.

Description. – Small, extremely involute, slightly compressed *Paranannites* with an ovoid whorl section, an arched venter, and convex flanks with maximum curvature near umbilicus, convergent to rounded venter. Umbilicus deep, but extremely narrow (not always visible), with oblique wall and rounded shoulders. Body chamber exceeds one whorl in length. Mature specimens exhibit obvious egressive coiling. No visible ornamentation. Suture line feebly ceratitic, very simple with only two saddles. Adult shell somewhat reminiscent of *Isculitoides* Spath, 1930 of Late Spathian age.

Discussion. – Although the morphology of this species is not entirely unlike the Paranannitidae, its suture line is very peculiar and resembles some of the simpler suture lines of the Melagathiceratidae. The validity of this assignment must be confirmed.

Paranannitidae gen. indet.

Pl. 33: 15a–d; Table 1

Occurrence. – Jin30; *Flemingites rursiradiatus* beds.

Description. – Moderately involute, slightly compressed shell with a low arched venter, indistinct ventral shoulders, and nearly parallel flanks. Umbilicus

moderately deep with perpendicular wall and rounded shoulders. Ornamentation consists only of very small plications. Suture line weakly ceratitic with two saddles.

Discussion. – This specimen is assigned to the Paranannitidae on the basis of its overall shape. It differs from *P.* aff. *aspenensis* by its quadratic whorl section and its more parallel flanks. It is also close to *P. cottreaui* Collignon, 1933–1934 from Madagascar. However, the specimen figured by Collignon is not well preserved. Thus, the attribution of the Chinese specimen to the Collignon's species needs more data.

Genus *Owenites* Hyatt & Smith, 1905

Type species. – *Owenites koeneni* Hyatt & Smith, 1905

Owenites koeneni Hyatt & Smith, 1905

Pl. 36: 1–8; Fig. 58

1905 *Owenites koeneni* sp. nov. Hyatt & Smith, p. 83, pl. 10: 1–22.
1915 *Owenites koeneni* – Diener, p. 214.
1932 *Owenites koeneni* – Smith, pl. 10: 1–22.
1932 *Owenites egrediens* Smith, p. 100, pl. 52: 6–8.
1932 *Owenites zitteli* Smith, p. 101, pl. 52: 1–5.
1934 *Owenites koeneni* – Spath, p. 185, fig. 57.
1947 *Owenites* aff. *egrediens* Kiparisova, p. 139, pl. 32: 1–3.
1955 *Kingites shimizui* Sakagami, p. 138, pl. 2: 2a–c.
1957 *Owenites koeneni* – Kummel, p. L138, fig. 171: 8a–c.
v 1959 *Owenites costatus* sp. nov. Chao, p. 249, pl. 22: 7–18, 22, 23, text-fig. 26c.
v 1959 *Owenites pakungensis* sp. nov. Chao, p. 248, pl. 21: 6–8.
v 1959 *Owenites pakungensis* var. *compressus* Chao, p. 248, pl. 21: 4, 5, text-fig. 26a.
v 1959 *Pseudowenites oxynotus* gen. et sp. nov. Chao, p. 252, pl. 23: 1–16, text-fig. 27a–d.
1959 *Owenites shimizui* Kummel, p. 430.
1960 *Owenites shimizui* Kummel & Sakagami, p. 6, pl. 2: 5–6.
1962 *Owenites koeneni* – Kummel & Steele, p. 674, pl. 101: 3–7.
1962 *Owenites koeneni* – Popov, p. 44, pl. 6: 6.
1965 *Owenites koeneni* – Kuenzi, p. 374, pl. 53: 1–6, text-figs 3d, 6.
1966 *Owenites koeneni* – Hada, p. 112, pl. 4: 2–4.
1968 *Owenites koeneni* – Kummel & Erben, p. 121: 12, pl. 19: 10–15.

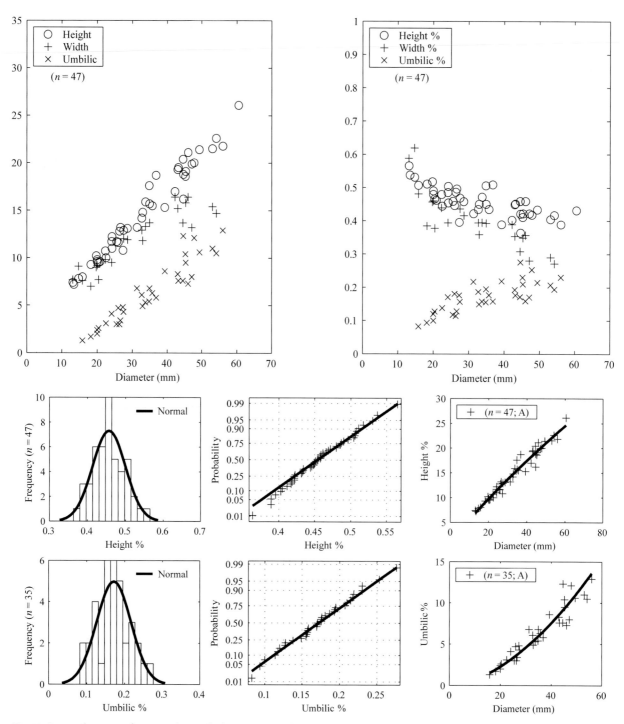

Fig. 58. Scatter diagrams of H, W and U, and of H/D, W/D and U/D, and histograms, probability plots of H/D, U/D and growth curves for *Owenites koeneni* (Jinya and Waili, *Owenites koeneni* beds). 'A' indicates allometric growth. 'I' indicates isometric growth.

1968 *Owenites carinatus* Shevyrev, p. 189, pl. 16: 1.
1968 *Owenites koeneni* – Zakharov, p. 94, pl. 18: 1–3.
1973 *Owenites koeneni* – Collignon, p. 139, pl. 4: 2, 39a.
1979 *Owenites koeneni* – Nichols & Silberling, pl. 1: 17–18.

1981 *Owenites koeneni* – Bando, p. 158, pl. 17: 7.
1984 *Owenites carinatus* Vu Khuc, p. 81, pl. 6: 1–4, text-fig. H16.
1984 *Pseudowenites oxynotus* Vu Khuc, p. 82, pl. 7: 3–4.
1990 *Owenites koeneni* – Shevyrev, p. 118, pl. 1: 5.
1995 *Owenites koeneni* – Shevyrev, p. 51, pl. 5: 1–3.

?2004 *Owenites pakungensis* Tong *et al.*, p. 199, pl. 2: 9–10, text-fig. 7.

Occurrence. – Jin12, 15, 18, 27, 42, 43, 44, 45, 46, 47, 99; NW1; T5, 8, 11; Yu3; *Owenites koeneni* beds.

Description. – Slightly involute, somewhat compressed shell with an inflated, lenticular whorl section and a typical, subangular to angular venter that may resemble a keel on mature specimens. Narrow, shallow umbilicus, becoming wider at maturity, with a low, steep wall and narrowly rounded shoulders. Coiling egressive. Surface generally smooth, but may exhibit weak, forward projected constrictions and folds as observed on *Paranannites*. Body chamber about one whorl in length. Suture line ceratitic with several divided umbilical lobes.

Discussion. – On the basis of their similar shell morphology, *O. egrediens* Smith (non-Welter) and *O. zitteli* Smith, 1932 were synonymized with *O. koeneni* by Kummel & Steele (1962) and Kummel & Erben (1968). *Pseudowenites oxynotus* was separated from *O. koeneni* by Chao (1959) because of slight variations in the auxiliary series of its suture line, but this cannot be justified if its ontogenetic development is considered (Kummel & Erben 1968). Kummel & Erben (1968) placed *O. costatus* Chao, 1959 in synonymy with *O. carpenteri* Smith, 1932, but measurements

indicate that this species cannot be distinguished from *O. koeneni*. Moreover, specimens of *O. costatus* figured by Chao (1959) exhibit a less oxyconic shape than the type species but are not well preserved on venter. Comparisons of well-preserved specimens from China and USA at different ontogenetic stages do not reveal any morphological differences. Whorl height and umbilical diameter of *O. koeneni* display significant allometric growth (Fig. 58).

Owenites simplex Welter, 1922

Pl. 35: 20–22; Fig. 59

1922 *Owenites simplex* nov. sp. Welter, p. 153, pl. 15: 1–8.
1934 *Parowenites simplex* – Spath, p. 187, fig. 58.
1959 *Owenites kwangsiensis* sp. nov. Chao, p. 250, pl. 22: 1–6, text-fig. 26b.
v 1959 *Owenites plicatus* sp. nov. Chao, p. 251, pl. 22: 19–21, 24–25, text-fig. 26e.
1968 *Owenites simplex* – Kummel & Erben, p. 122, fig. 12k, n–o.

Occurrence. – Jin45; Yu1; *Owenites koeneni* beds.

Description. – Slightly involute shell, similar to *O. koeneni*, but more compressed, with a subangular to

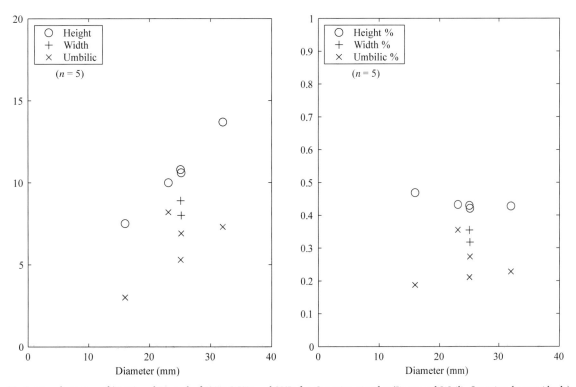

Fig. 59. Scatter diagrams of H, W and U, and of H/D, W/D and U/D for *Owenites simplex* (Jinya and Waili, *Owenites koeneni* beds).

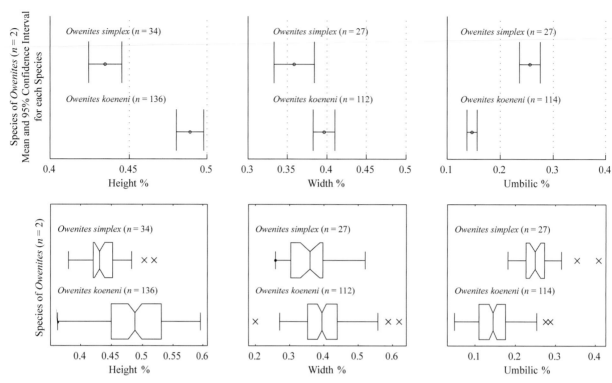

Fig. 60. Mean and box plots for two species of *Owenites* found in the Luolou Formation.

angular venter, bearing a very weak keel. Narrow, moderately deep umbilicus with a low, perpendicular wall and rounded shoulders. Ornamentation consists of conspicuous, prorsiradiate, sigmoidal ribs, as well as a few, very small, fold-like plications on umbilical shoulder. Suture line first presented as goniatitic (see Kummel & Erben 1968), but it is ceratitic and similar to *O. koeneni*.

Discussion. – *O. simplex* is easily distinguished from *O. koeneni* by its less involute coiling and its more compressed shell (Fig. 60).

Owenites carpenteri Smith, 1932

Pl. 43: 15–16; Table 1

1932 *Owenites carpenteri* n. sp. Smith, p. 100, pl. 54: 31–34.
1966 *Owenites carpenteri* – Hada, p. 112, pl. 4: 1a–e.
1968 *Owenites carpenteri* – Kummel & Erben, p. 122, fig. 12l.
1973 *Owenites carpenteri* – Collignon, p. 139, pl. 4: 5, 5a.

Occurrence. – Jin47; T12; *Owenites koeneni* beds.

Description. – Extremely involute, compressed shell with a slightly inflated, lenticular whorl section, a very narrowly rounded to subangular venter, convex flanks, and an occluded umbilicus. Ornamentation consists of thin, slightly projected growth lines and a few folds. Suture line similar to *O. koeneni*.

Discussion. – Apparently, our specimens are less ornamented than the American specimens described by Smith (1932). However, the occluded umbilicus and narrowly curved venter of our specimens are diagnostic of this species.

Superfamily Sagerataceae Hyatt, 1884

Family Hedenstroemiidae Hyatt, 1884

Genus *Pseudosageceras* Diener, 1895

Type species. – *Pseudosageceras* sp. indet. Diener, 1895.

Pseudosageceras multilobatum Noetling, 1905

Pl. 37: 1–5; Fig. 61

1905 *Pseudosageceras multilobatum* Noetling, pl. 25: 1a–b; pl. 26: 3a–b.
1905 *Pseudosageceras intermontanum* sp. nov. Hyatt & Smith, p. 99, pl. 4: 1–3; pl. 5: 1–6; pl. 63: 1–2.
1909 *Pseudosageceras multilobatum* – Krafft & Diener, p. 145, pl. 21: 5.

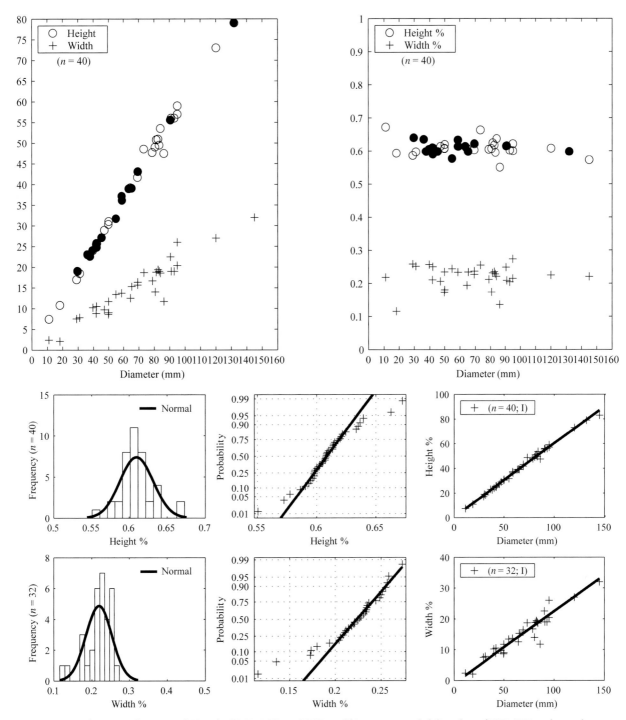

Fig. 61. Scatter diagrams of H, W, and U, and of H/D, W/D and U/D, and histograms, probability plots of H/D, U/D and growth curves for *Pseudosageceras multilobatum* (solid circles indicate specimens from Jinya and Waili, *Flemingites rursiradiatus* beds; open circles indicate specimens from Siberia, Caucasus, Madagascar, Idaho; data from Dagys & Ermakova 1990; Shevyrev 1968; Collignon 1933–1934; Kummel & Steele 1962, respectively).

1911 *Pseudosageceras multilobatum* – Wanner, p. 181, pl. 7: 4.

1911 *Pseudosageceras drinense* Arthaber, p. 201, pl. 17: 6–7.

1922 *Pseudosageceras multilobatum* – Welter, p. 94: 3.

1929 *Pseudosageceras intermontanum* Mathews, p. 3, pl. 1: 18–22.

1932 *Pseudosageceras multilobatum* – Smith, p. 87–89, pl. 4: 1–3; pl. 5: 1–6; pl. 25: 7–16; pl. 60: 32; pl. 63: 1–6.

1934 *Pseudosageceras multilobatum* – Collignon, pp. 56–58, pl. 11: 2.

1934 *Pseudosageceras multilobatum* – Spath, p. 54, fig. 6a.

1947 *Pseudosageceras multilobatum* – Kiparisova, p. 127, pl. 25: 3–4.

1947 *Pseudosageceras multilobatum* var. *giganteum* Kiparisova, p. 127, pl. 26: 2–5.

1948 *Pseudosageceras* cf. *clavisellatum* Renz & Renz, p. 90, pl. 16: 3.

1948 *Pseudosageceras drinense* Renz & Renz, p. 92, pl. 16: 6.

1948 *Pseudosageceras intermontanum* Renz & Renz, pp. 90–92, pl. 16: 4, 7.

v 1959 *Pseudosageceras multilobatum* – Chao, p. 183, pl. 1: 9, 12.

v 1959 *Pseudosageceras curvatum* sp. nov. Chao, p. 185, pl. 1: 13, 14, text-fig. 5a.

v 1959 *Pseudosageceras tsotengense* sp. nov. Chao, p. 184, pl. 1: 7, 8, text-fig. 5b.

?1959 *Pseudosageceras multilobatum* var. nov. Jeannet, p. 30, pl. 6: 1.

1961 *Pseudosageceras schamarense* Kiparisova, p. 31, pl. 7: 3–4.

1961 *Pseudosageceras multilobatum* var. *gigantea* Popov, p. 13, pl. 2: 1–2.

non 1962 *Pseudosageceras multilobatum* – Kummel & Steele, p. 701, pl. 102: 1–2.

?1966 *Pseudosageceras multilobatum* – Hada, p. 112, pl. 4: 6.

1968 *Pseudosageceras multilobatum* – Kummel & Erben, p. 112, pl. 19: 9.

1968 *Pseudosageceras multilobatum* – Shevyrev, p. 791, pl. 1: 1–2.

?1973 *Pseudosageceras multilobatum* – Collignon, p. 5, pl. 1: 1.

1978 *Pseudosageceras multilobatum* – Weitschat & Lehmann, p. 95, pl. 10: 2a–b.

1984 *Pseudosageceras multilobatum* – Vu Khuc, p. 26, pl. 1: 1.

1994 *Pseudosageceras multilobatum* – Tozer, p. 83, pl. 18: 1a–b; p. 384: 17.

Occurrence. – Jin4, 11, 13, 28, 29, 30, 51; WFB; FW2, 3, 4, 5; Sha1; *Flemingites rursiradiatus* beds. Jin27, 44; *Owenites koeneni* beds.

Description. – Extremely involute, compressed oxycone, with an occluded umbilicus, a very narrow, concave, bicarinate venter, especially on mature specimens, and weakly convex flanks, convergent from umbilicus to venter. Surface smooth without ornamentation. Suture line ceratitic, complex and composed of many adventitious elements with characteristic trifid lateral lobe. Other lobes are bifid.

Discussion. – *P. multilobatum* is one of the most cosmopolitan and long-ranging species of the Early Triassic, and has its acme in the Smithian stage.

Genus *Clypites* Waagen, 1895

Type species. – *Clypites typicus* Waagen, 1895

Clypites sp. indet.

Pl. 38: 1–4; Fig. 62

Occurrence. – Jin62; *Clypites* sp. indet. beds.

Description. – Compressed oxycone with an occluded umbilicus, a subtabulate venter (becoming narrowly rounded on largest specimens), angular ventral shoulders, and nearly flat, but slightly convex and convergent flanks. Shell smooth, with prorsiradiate, convex to biconcave growth lines. Suture line agreeing in plan with that of Hedenstroemiidae. Although the detailed shape of the ventral part of the suture line is not known, there is not much space in which the adventitious lobe(s) can be accomodated, thus suggesting the presence of a single adventitious lobe only. Auxiliary lobes and saddles typically bifid.

Discussion. – Among hedenstroemids, *Clypites* has a similar shell shape and a very close suture line characterized by a narrow adventitious lobe and bifid auxiliary lobes. *C. kingianus* Waagen is the closest representative of *Clypites*. *Clypites* sp. indet. from Waili is easily distinguished from involute and compressed protychitids (e.g. *Clypeoceras*) by its convex to biconcave growth striae, its occluded umbilicus and distinctive auxiliary lobes. The previously know stratigraphical distribution of *Clypites* is restricted to the Ceratite Marls (Salt Range), which exclusively contain faunas of Dienerian age. Hence, the *Clypites* sp. indet. beds from northwestern Guangxi are assigned to the Dienerian.

Genus *Hedenstroemia* Waagen, 1895

Type species. – *Ceratites hedenstroemi* Keyserling, 1845

Hedenstroemia augusta n. sp.

Pl. 39: 1–11; Fig. 63

Diagnosis. – Hedenstroemiidae with extremely involute coiling, a tabulate venter and flanks with two different angles of slope on juvenile specimens.

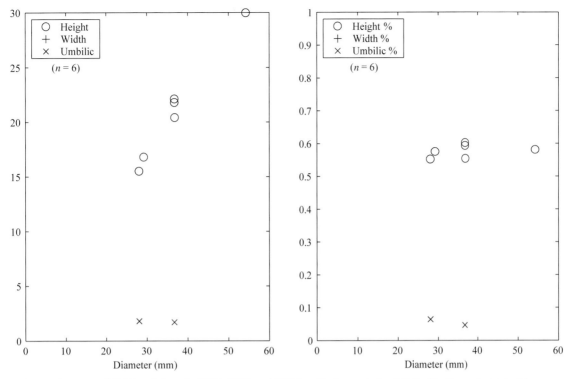

Fig. 62. Scatter diagrams of H, W and U, and of H/D, W/D and U/D for *Clypites* sp. indet. (Waili, *Clypites* sp. indet. beds).

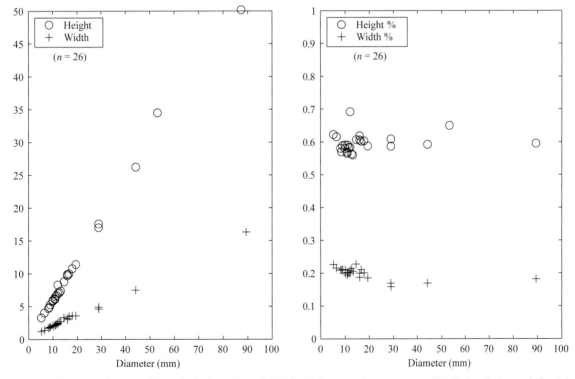

Fig. 63. Scatter diagrams of H, W and U, and of H/D, W/D and U/D for *Hedenstroemia augusta* n. sp. (Waili, *Anasibirites multiformis* beds).

Holotype. – PIMUZ 26138, Loc. NW13, Waili, *Anasibirites multiformis* beds, Smithian.

Derivation of name. – From the Latin word: augusta, meaning 'emperor's wife'.

Occurrence. – Jin33, 90, 91, 105, 106; FW7, 12; NW13, 15; Yu5, 6; *Anasibirites multiformis* beds.

Description. – Extremely involute, compressed oxycone with an occluded umbilicus, a narrow, weakly

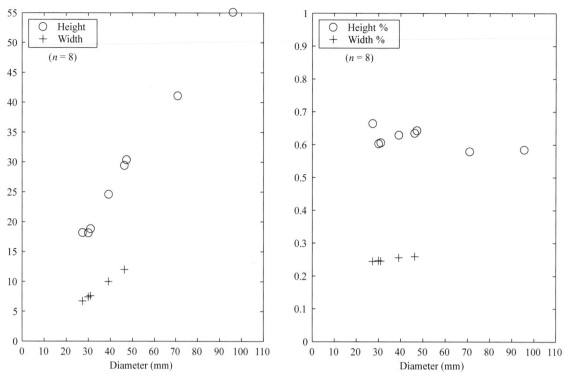

Fig. 64. Scatter diagrams of H, W, and U, and of H/D, W/D and U/D for *Cordillerites antrum* n. sp. (Waili, *Kashmirites kapila* beds).

bicarinate venter (tabulate on internal mold), and flanks on juvenile specimens with weak, but distinct longitudinal line at about mid-flank, marking a very slight change in slope between umbilical and ventral portions of flank. Umbilical portion nearly flat, ventral portion slightly convergent to narrow venter. On larger specimens, this change of slope angle disappears and flanks become slightly convex. Ornamentation smooth with only thin, but conspicuous, sinuous growth lines. Concave part of growth line located on ventral half of flank. Suture line typical of Hedenstroemiidae with complex architecture exhibiting numerous saddles and a very long auxiliary series. Lateral lobe displays many indentations, thus differentiating this species from other similarly shaped ammonoids, e.g. *P. multilobatum*.

Discussion. – The occurrence of this new species of *Hedenstroemia* at the very end of the Smithian clearly documents that this genus is long ranging. *H. augusta* n. sp. essentially differs from *H. hedenstroemi* by the presence of a longitudinal line at about mid-flank on juvenile specimens. The whorl width of *H. augusta* n. sp. is somewhat less than that of *Pseudosageceras* and *Cordillerites* (see Fig. 65).

Genus *Cordillerites* Hyatt & Smith, 1905

Type species. – *Cordillerites angulatus* Hyatt & Smith, 1905

Cordillerites antrum n. sp.

Pl. 40: 1–9; Fig. 64

Diagnosis. – Hedenstroemid with a distinctive tabulate venter, slightly convex flanks, sinuous plications and suture line similar to *Pseudosageceras*, but with a trifid lateral and a trifid first umbilical lobes.

Holotype. – PIMUZ 26148, Loc. Jin61, Jinya, *Kashmirites kapila* beds, Smithian.

Derivation of name. – From the Latin word: antrum, meaning hollow cave.

Occurrence. – Jin61, 64, 65, 66; *Kashmirites kapila* beds.

Description. – Extremely involute, compressed oxycone with an occluded umbilicus and a concave, bicarinate venter on smaller specimens, becoming subtabulate to slightly rounded on body chamber of largest specimens. Flanks nearly flat and convergent with very slight convex curvature, forming somewhat ovoid whorl section on largest specimens. Ornamentation smooth with faint, very thin, sinuous growth lines, not particularly conspicuous on largest specimens. Small plications visible near venter on some specimens. Suture line exhibits two adventitious lobes. The

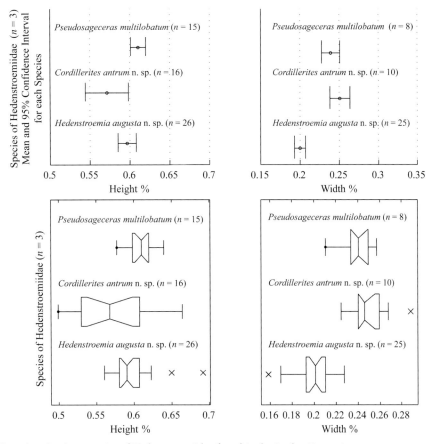

Fig. 65. Mean and box plots for three species of Hedenstroemiidae found in the Luolou Formation.

two following lobes are trifid as is lateral lobe of *Pseudosageceras.*

Discussion. – This genus clearly belongs to the Hedenstroemiidae based on its morphology, suture line and diagnostic measurements (Fig. 65). *Cordillerites* possesses a combination of several characters exhibited by different genera of Hedenstroemiidae, thus making it very difficult to identify without the aid of its suture line. Although it is similar to that of *Pseudosageceras,* its suture line differs by its less complicated structure and characteristic first umbilical trifid lobe. For juvenile specimens, auxiliary elements are less numerous and complex than *Pseudosageceras. Clypites* and *Tellerites* are much different from *Cordillerites* in that their suture lines do not display so many adventious elements.

Cordillerites antrum n. sp. appears to be more compressed than *C. angulatus,* and it also exhibits sinuous plications and striae not present on the type species.

Genus *Mesohedenstroemia* Chao, 1959

Type species. – *Mesohedenstroemia kwangsiana* Chao, 1959

Mesohedenstroemia kwangsiana Chao, 1959

Pl. 41: 1–8; Fig. 66

?1895 *Parakymatites discoides* n. gen. et sp. Waagen, p. 214, pl. 36: 3.

v 1959 *Mesohedenstroemia kwangsiana* gen. et sp. nov. Chao, p. 266, pl. 34: 1–18, text-fig. 33b–d.

v 1959 *Mesohedenstroemia inflata* sp. nov. Chao, p. 267, pl. 35: 4–8, text-fig. 33a.

Occurrence. – Jin4, 11, 13, 23, 24, 28, 29, 30, 41, 51; FW4/5; Sha1; T6, T50; *Flemingites rursiradiatus* beds. Jin10; *Owenites koeneni* beds.

Description. – Very involute, compressed, discoidal shell with a broad, distinctive tabulate venter, angular ventral shoulders, and flat or gently convex flanks. Umbilicus generally very small on juvenile specimens, opening somewhat on body chamber, may even be wider on largest specimens. When open, umbilicus is moderately deep with perpendicular wall and abruptly rounded shoulders. Shell usually smooth, but can exhibit growth lines on flanks, curving slightly forward near venter. Suture line exhibits one

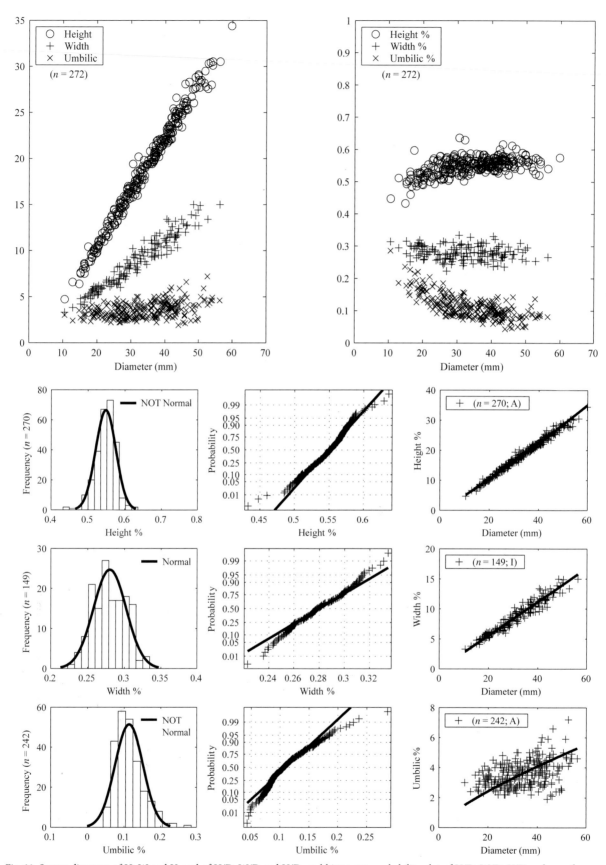

Fig. 66. Scatter diagrams of H, W and U, and of H/D, W/D and U/D, and histogram, probability plot of H/D, W/D, U/D and growth curve for *Mesohedenstroemia kwangsiana* (Jinya, *Flemingites rursiradiatus* beds). 'A' indicates allometric growth. 'I' indicates isometric growth.

adventitious lobe, but not with trifid division as seen in some genera of the family.

Discussion. – Whorl height and umbilical diameter exhibit allometric growth while whorl width displays isometric growth (Fig. 66). *Mesohedenstroemia* is similar to *Hedenstroemia*, but it is characterized by a wider, distinctive tabulate venter and a simple suture line without adventious elements. Kummel & Steele (1962) consider *Mesohedenstroemia* to be a synonym of *Pseudohedenstroemia*, but the latter is closer to *Hedenstroemia*. *Lingyunites* Chao, 1950 with its discoidal whorl section, tabulate venter and occluded umbilicus is closely related to *Mesohedenstroemia*, but its suture line is simpler. It can be difficult to distinguish between these two genera, especially juvenile specimens, without the aid of a well-preserved suture line and diagnostic measurements. The absence of adventious elements in the suture line of *Mesohedenstroemia* questions its attribution to Hedenstroemiidae. This attribution is mainly based on the complex auxiliary series and number of elements of its suture line.

Mesohedenstroemia planata Chao, 1959

Pl. 41: 9a–c; Table 1

v 1959 *Mesohedenstroemia planata* sp. nov. Chao, p. 268, pl. 35: 1–3, text-fig. 33e.

Occurrence. – Jin45; *Owenites koeneni* beds.

Description. – Extremely involute, very compressed, discoidal shell with an occluded umbilicus, a sub-tabulate venter, abruptly rounded ventral shoulders, and nearly parallel flanks. No visible ornamentation. Suture line simple, ceratitic, similar to *M. kwangsiana*.

Discussion. – The single specimen illustrated as *M. planata* by Chao (1959) appears to have a somewhat questionable umbilicus that may have been the result of excessive preparation. Other diagnostic characters seem to be similar to those of our only specimen. *M. planata* essentially differs from *M. kwangsiana* by its extreme involute coiling.

Family Aspenitidae (Spath, 1934)

Genus *Aspenites* Hyatt & Smith, 1905

Type species. – *Aspenites acutus* Hyatt & Smith, 1905

Aspenites acutus Hyatt & Smith, 1905

Pl. 42: 1–9; Fig. 67

```
   1905 Aspenites acutus sp. nov. Hyatt & Smith,
        p. 96, pl. 2: 9–13; pl. 3: 1–5.
  ?1909 Hedenstroemia acuta – Krafft & Diener,
        p. 157, pl. 9: 2a–d.
   1915 Aspenites acutus – Diener, p. 59, fig. 20.
   1922 Aspenites acutus – Welter, p. 98, fig. 7.
   1922 Aspenites laevis nov. sp. Welter, p. 99, pl. 1:
        4–5.
   1932 Aspenites acutus – Smith, p. 86, pl. 2: 9–13;
        pl. 3: 1–5; pl. 30: 1–26; pl. 60: 4–6.
   1932 Aspenites laevis Smith, p. 86, pl. 28: 28–33.
   1932 Aspenites obtusus Smith, p. 86, pl. 31: 8–10.
   1934 Aspenites acutus – Spath, p. 229, fig. 76.
  ?1934 Parahedenstroemia acuta – Spath, p. 221, fig. 70.
   1957 Aspenites acutus – Kummel, p. L142,
        fig. 173: 1a–c.
 v 1959 Aspenites acutus – Chao, p. 269, pl. 35: 12–
        18, 23, text-fig. 34a.
 v 1959 Aspenites laevis Chao, p. 270, pl. 35: 9–11,
        23, text-fig. 34b.
   1962 Aspenites acutus – Kummel & Steele, p. 692,
        pl. 99: 16–17.
   1962 Hemiaspenites obtusus Kummel & Steele,
        p. 666, pl. 99: 18.
   1979 Aspenites? cf. acutus – Nichols & Silberling,
        pl. 1: 10–11.
   1979 Aspenites acutus – Nichols & Silberling, pl. 1:
        12–14.
```

Occurrence. – Jin4, 23, 28, 29, 30, 41, 51; Sha1; T6, T50; *Flemingites rursiradiatus* beds. Jin10, 27; *Owenites koeneni* beds.

Description. – Extremely involute, very compressed oxycone with an acutely keeled venter, an occluded umbilicus and slightly convex flanks with maximum curvature at mid-flank. Umbilical region slightly depressed. Surface generally smooth or ornamented with falcoid growth lines, and radial folds disappearing near venter. Suture line complex with wide, curved series of small auxiliary saddles.

Discussion. – *Aspenites acutus* is easily distinguishable among the Aspenitidae by its occluded umbilicus and very acute venter. Whorl height also exhibits isometric growth (Fig. 67). Prior to our study, the genus consisted of only two species: *A. acutus* Hyatt & Smith, 1905 and *A. laevis* Welter, 1922. Kummel & Steele (1962) differentiated between these two species based only on their assertion that the suture line of *A. laevis* was more complex. However, all comparisons and especially diagnostic measurements lead us to

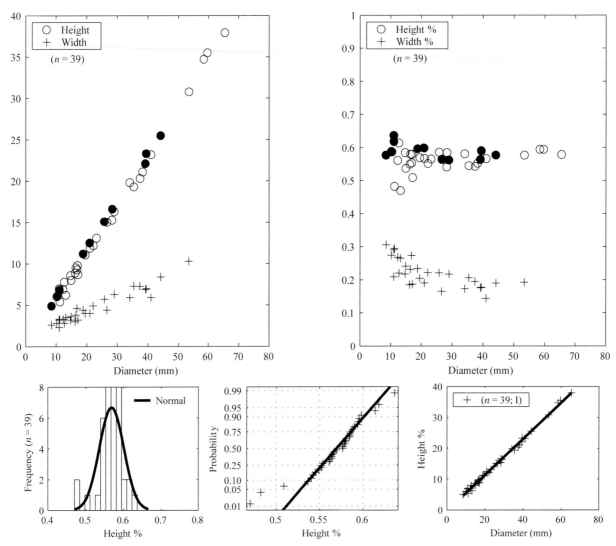

Fig. 67. Scatter diagrams of H, W and U, and of H/D, W/D and U/D, and histogram, probability plot of H/D, and growth curve for *Aspenites acutus* (Jinya and Yuping, *Flemingites rursiradiatus* beds; solid circles indicate specimens from Kummel & Steele 1962 given for comparison). 'A' indicates allometric growth. 'I' indicates isometric growth.

conclude they should be synonymized within a single species. Undeniably, the suture line of the *A. laevis* type specimen represents the adult stage (see Kummel & Steele 1962 for comparison).

Similarly, the genus *Hemiaspenites* Kummel & Steele, 1962 must also be synonymized with *Aspenites*. Indeed, Kummel & Steele stressed that the suture line of *Hemiaspenites* was different, but their illustrations (text-fig. 5f–g) clearly indicate the suture lines, either were excessively ground in preparation, or are poorly preserved, and actually represent a juvenile stage of *A. acutus*.

?*Aspenites* sp. indet.

Pl. 42: 10–11

Occurrence. – Jin27, 45, 99; NW1; Yu1; *Owenites koeneni* beds.

Description. – Extremely involute, very compressed oxycone similar to *A. acutus*, but much larger with an acutely keeled venter on largest specimens and flanks slightly more convex than *A. acutus*. Character of umbilicus unknown. Our specimens consist only of adult body chambers. Umbilical area relatively shallow in comparison with large size of our specimens. No visible umbilical shoulders. Only some sinuous growth lines and folds are visible on shell. Suture line unknown, all specimens are adult body chambers.

Discussion. – Material here referred to as ?*Aspenites* sp. indet. may represent adult body chamber of

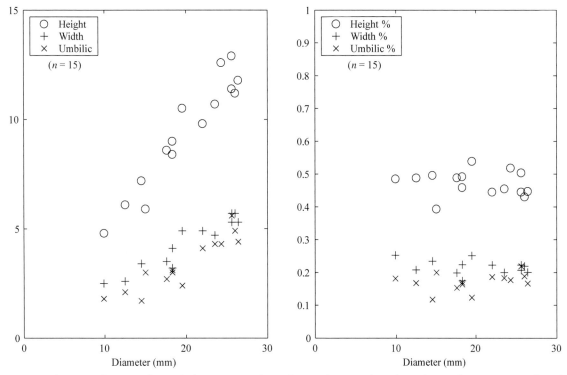

Fig. 68. Scatter diagrams of H, W, and U, and of H/D, W/D and U/D for *Pseudaspenites layeriformis* (Jinya, *Flemingites rursiradiatus* beds).

A. acutus. However, given the fragmentary nature of these specimens and the absence of suture line, a more precise identification remains impossible.

Genus *Pseudaspenites* Spath, 1934

Type species. – *Aspenites layeriformis* Welter, 1922

Pseudaspenites layeriformis (Welter, 1922)

Pl. 43: 1–6; Fig. 68

p 1922 *Aspenites layeriformis* nov. sp. Welter, p. 97 (*pars*), pl. 1: 6–7 (fig. 8 = ? *Aspenites acutus*).
  1934 *Pseudaspenites layeriformis* – Spath, p. 230, fig. 77.
v 1959 *Inyoites striatus* sp. nov. Chao, p. 197, pl. 2: 22–26.
v 1959 *Inyoites obliplicatus* sp. nov. Chao, p. 198, pl. 2: 7, 17–21, 27.

Occurrence. – Jin4, 13, 23, 28, 29, 30; FW2, 3, 4, 5; Sha1; T6, T50; *Flemingites rursiradiatus* beds.

Description. – Involute, very compressed oxycone with a relatively broad umbilicus, an acutely keeled venter and generally flat, nearly smooth, convergent flanks. Extremely shallow umbilicus with slightly angular, very short shoulders. Ornamentation on some specimens consists of extremely fine, forward projecting, sigmoidal striation that forms a distinctive keel. On some specimens, these projections form a delicate, crenulated keel. Suture line ceratitic with smaller auxiliary series than *A. acutus.* Lobes also broader and more indented.

Discussion. – The formation of a crenulated keel on certain specimens is puzzling when compared to other, similar sized specimens without this unusual ornamentation. One obvious possibility is that specimens with the crenulated keel represent a different, but very rare species. However, their diagnostic measurements are all consistent with *A. layeriformis.*

The illustration of the suture line of the type species by Welter (1922) strongly resembles that of *Aspenites.* The suture line of *Pseudaspenites layeriformis* has broad, well-indented lobes, reduced auxiliary series and lack of adventious elements. These differences suggest the possibility of confusion in Welter's illustration of the suture lines for these two genera.

Pseudaspenites evolutus n. sp.

Pl. 43: 7–11; Fig. 69

Diagnosis. – *Pseudaspenites* with more evolute coiling than *P. layeriformis* and with a lower keel.

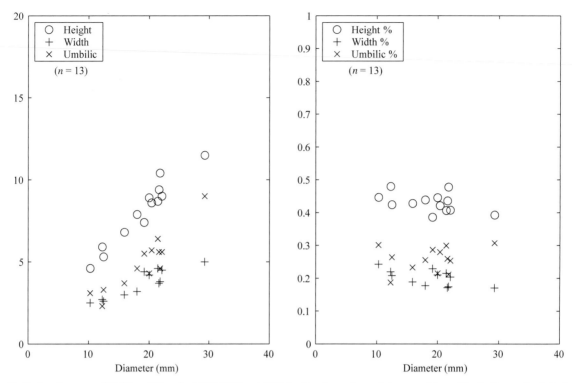

Fig. 69. Scatter diagrams of H, W, and U, and of H/D, W/D and U/D for *Pseudaspenites evolutus* n. sp. (Jinya, *Flemingites rursiradiatus* beds).

Holotype. – PIMUZ 26186, Loc. Jin30, Jinya, *Flemingites rursiradiatus* beds, Smithian.

Derivation of name. – Named for its evolute coiling.

Occurrence. – Jin4, 23, 28, 29, 30; Sha1; *Flemingites rursiradiatus* beds.

Description. – Whorl section nearly identical to *P. layeriformis*, but less involute, with a more acute venter (with a thin keel) and a relatively broad umbilicus. Coiling slightly egressive on largest specimens. Weak folds and growth lines are visible, but do not extend onto the keel as in *P. layeriformis*. Suture line unknown.

Discussion. – *P. evolutus* n. sp. essentially differs from *P. layeriformis* by its more evolute coiling, but this distinction is often tenuous, especially for small-sized specimens.

Pseudaspenites tenuis (Chao, 1959)

Pl. 43: 12–14; Fig. 70

v 1959 *Aspenites tenuis* sp. nov. Chao, p. 271, pl. 35: 19–22, text-fig. 34c.

Occurrence. – Jin4, 10, 23, 30; *Flemingites rursiradiatus* beds and *Owenites koeneni* beds.

Description. – Whorl section similar to *P. layeriformis*, but much more compressed. Degree of involution similar to *P. layeriformis* for small specimens. Ornamentation consists only of very weak folds. Suture line typical of *Pseudaspenites* with broad indented lobes and very small auxiliary series.

Discussion. – *P. tenuis* is much more compressed than all other species of *Pseudaspenites* (see Fig. 71). It also lacks a keel.

Superfamily Otocerataceae Hyatt, 1900

Family Anderssonoceratidae Ruzhencev, 1959

Genus *Proharpoceras* Chao, 1950

Type species. – *Proharpoceras carinatitabulatum* Chao, 1950

Proharpoceras carinatitabulatum Chao, 1950

Pl. 38: 5–9; Fig. 72

v 1950 *Proharpoceras carinatitabulatus* gen. et sp. nov. Chao, p. 8, pl. 1: 8a–b.
v 1950 *Tuyangites marginalis* Chao, p. 9, pl. 1: 9.
v 1959 *Proharpoceras carinatitabulatum* – Chao, p. 324, pl. 43: 1–7.

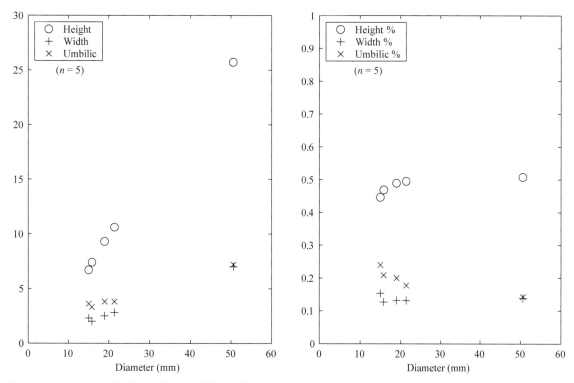

Fig. 70. Scatter diagrams of H, W and U, and of H/D, W/D and U/D for *Pseudaspenites tenuis* (Jinya, *Flemingites rursiradiatus* beds).

v 1959   *Tuyangites marginalis* Chao, p. 327, pl. 43: 17–18.

v 1959   *Prosphingites sinensis* sp. nov. Chao, p. 297, number 12583.

1968   *Proharpoceras carinatitabulatum* – Zakharov, p. 147, pl. 29: 6.

v 2007a   *Proharpoceras carinatitabulatum* – Brayard *et al.*, fig. 3a–s, y–z.

Occurrence. – Jin45; Yu1; *Owenites koeneni* beds.

Description. – Moderately evolute, thick platycone with a quadratic whorl section, distinctive, 'tabulate' to tectiform venter with a raised keel, and parallel, nearly flat or gently convex flanks. On specimens with well-preserved outer shell, venter is seen to actually be tricarinate with very small marginal keels rising from ventral shoulders. Umbilicus wide, with moderately high, slightly oblique wall and rounded shoulders. Body chamber ca. one whorl in length. Ornamentation consists of ridges on flanks, then becoming strongly projected on venter. Convex growth lines are visible. Projected, ventral growth lines converge at central keel on largest specimens. Suture line composed of narrow ventral lobe and two broad, lateral ceratitic lobes. Saddles asymmetrical. Suture line first described as goniatitic by Chao (1959), and then as ceratitic by Zakharov (1968).

Discussion. – The distinctive tricarinate morphology of this species is strongly reminiscent of the Permian Anderssonoceratidae and Araxoceratidae, with the exception of its umbilical shoulders, which are not raised. Its suture line is also similar to the Late Permian Anderssonoceratidae. These characteristics justify its revised systematic placement within this family (Brayard *et al.* 2007a).

Tuyangites, also described by Chao (1950), exhibits identical morphological characters (especially backward projected ribs on its flanks and marginal ridges), but it differs by having weak nodes on the inner whorls. These inner nodes can be observed not only on the holotype (number 12278) in Chao's collection at Nanjing, but also on some of our *Proharpoceras* specimens. Therefore, the near identical morphology and ornamentation of *Tuyangites* provide ample justification to synonymize these two genera. Chao (1959) also described a second species, *Proharpoceras tricarinatum*, which could not be duplicated here.

Incertae Sedis

Genus *Larenites* n. gen.

Type species. – *Flemingites reticulatus* Tozer, 1994

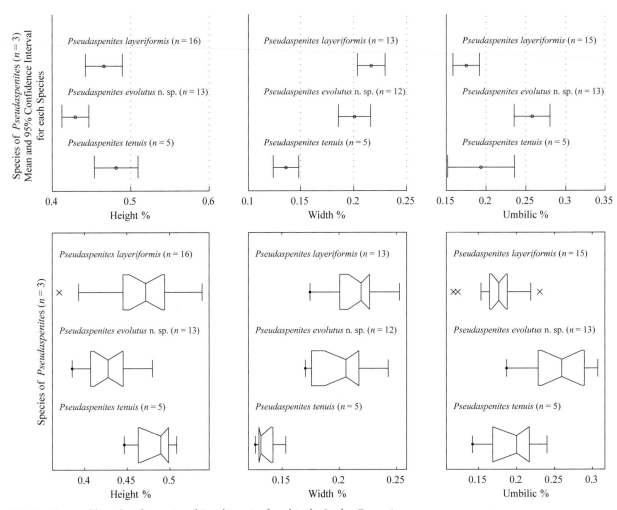

Fig. 71. Mean and box plots for species of *Pseudaspenites* found in the Luolou Formation.

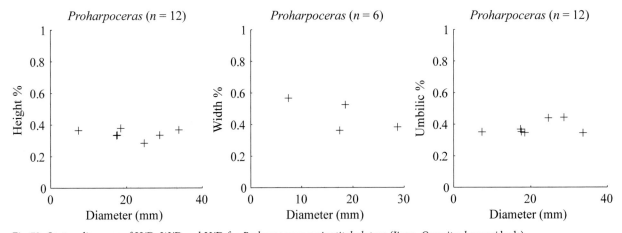

Fig. 72. Scatter diagrams of H/D, W/D and U/D for *Proharpoceras carinatitabulatum* (Jinya, *Owenites koeneni* beds).

Composition of the genus. – *Larenites reticulatus* (Tozer, 1994).

Diagnosis. – Very involute with a broad venter, plications and strigation.

Derivation of name. – Named after the small village of Laren (Guangxi Province, south China).

Occurrence. – *Kashmirites kapila* beds and *Flemingites rursiradiatus* beds.

Discussion. – This species was provisionally assigned to *Flemingites* by Tozer (1994), but it is in fact rather different from the type species of *Flemingites*. Therefore, a new genus is erected and the species name is retained. This genus, with its involute coiling, also resembles *Subflemingites* Spath, 1934, but the latter is smooth. Assignment of *Larenites reticulatus* to Flemingitidae by Tozer remains uncertain in the absence of illustration of the suture line. Its general shape strongly suggests affinity with Proptychitidae.

Tozer (1994) reported the occurrence of this species from the Late Dienerian Sverdrupi Zone of British Columbia. The new occurrences from Guangxi come from the *Kashmirites kapila* beds and *Flemingites rursiradiatus* beds, thus expanding the range of this species across the Dienerian–Smithian boundary.

Larenites cf. *reticulatus* (Tozer, 1994)

Pl. 24: 1–2

1994 *Flemingites reticulatus* n. sp. Tozer, p. 71, pl. 20: 5–7

Occurrence. – Jin23, 66; *Kashmirites kapila* beds and *Flemingites rursiradiatus* beds.

Description. – Involute, thick platycone with a subtabulate venter, rounded ventral shoulders and convex flanks with maximum lateral curvature near mid-flank. Umbilicus with high, slightly overhanging wall and rounded shoulders. Ornamentation consists of conspicuous folds and irregular, sinuous plications as well as obvious strigation on venter and a portion of ventral shoulders. Suture line unknown for our specimens.

Genus *Guodunites* n. gen.

Type species. – *Guodunites monneti* n. sp.

Composition of the genus. – Type species only.

Diagnosis. – Involute, compressed platycone with dense, thickened growth striae, resembling radial lirae of some Phylloceratidae. Suture line subammonitic with a markedly indented ventral saddle.

Derivation of name. – Named after Kuang Guodun (Guangxi Bureau of Geology and Mineral Resources, Nanning).

Occurrence. – *Owenites koeneni* beds.

Discussion. – This genus embodies a unique combination of outer shell shape, striation and suture line, which so far, is unknown in the Early Triassic, and consequently, family assignment is not possible.

Guodunites monneti n. sp.

Pl. 44: 1–2; Table 1

Diagnosis. – As for the genus.

Holotype. – PIMUZ 26193, Loc. Jin99, Jinya, *Owenites koeneni* beds, Smithian.

Derivation of name. – Named after Claude Monnet (PIMUZ).

Occurrence. – Jin12, 99; Yu7; *Owenites koeneni* beds.

Description. – Involute, compressed platycone with a broadly arched to subtabulate venter, becoming more highly arched on larger specimens, and slightly convex flanks with maximum curvature near venter. Umbilicus relatively shallow for juvenile specimens, but umbilical characteristics unknown for largest specimens due to fragmentary nature of material. Ornamentation on large specimens consists of very fine, but conspicuous radial lirae. A few plications are visible on smaller specimens. Suture line subammonitic, but incompletely known. It exhibits a high, first lateral saddle, and a large ventral saddle, which is crenulated on its ventral side.

Genus *Procurvoceratites* n. gen.

Type species. – *Procurvoceratites pygmaeus* n. sp.

Composition of the genus. – Three species: *Procurvoceratites pygmaeus* n. sp., *P. ampliatus* n. sp. and *P. tabulatus* n. sp.

Diagnosis. – Involute, very small platycone with forward projected constrictions.

Derivation of name. – From the Latin word: procurvus, meaning 'bent forward'.

Occurrence. – *Flemingites rursiradiatus* beds.

Discussion. – This genus is characterized by its distinctive, forward projected constrictions, which resemble the ornamentation of some Melagathiceratidae (e.g. *Juvenites*). However, its extremely small size, and high projection angle of its constrictions are very unique characteristics. Since its suture

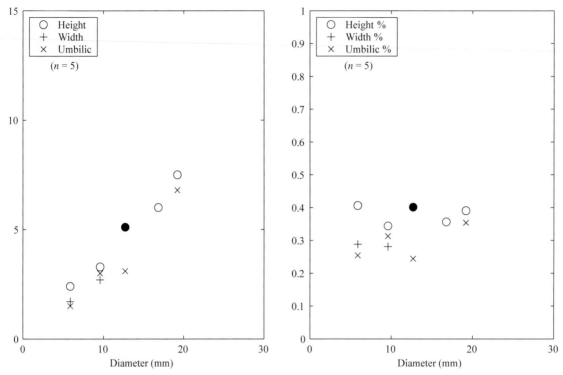

Fig. 73. Scatter diagrams of H, W, and U, and of H/D, W/D and U/D for *Procurvoceratites pygmaeus* n. gen., n. sp. and *P. ampliatus* n. gen., n. sp. represented as a solid circle (Jinya, *Flemingites rursiradiatus* beds).

line is unknown, it cannot be assigned to a specific family.

Procurvoceratites pygmaeus n. sp.

Pl. 44: 3–5; Fig. 73

Diagnosis. – Very small shell with involute coiling and sinuous projections.

Holotype. – PIMUZ 26195, Loc. Jin28, Jinya, *Flemingites rursiradiatus* beds, Smithian.

Derivation of name. – From the Latin word: pygmaeus, referring to the extremely small size of its adult stage.

Occurrence. – Jin4, 28, 29, 30; *Flemingites rursiradiatus* beds.

Description. – Moderately involute, very small-sized platycone with a circular venter, rounded ventral shoulders, and concave flanks. Umbilicus with short wall and rounded shoulders. Ornamentation consists only of strongly projected, prorsiradiate constrictions. Maturity is reached at a diameter of about 8 mm. Suture line unknown.

Procurvoceratites ampliatus n. sp.

Pl. 44: 6; Fig. 73

Diagnosis. – *Procurvoceratites* with forward projected, concave constrictions and a thick whorl width.

Holotype. – PIMUZ 26198, Loc. Jin30, Jinya, *Flemingites rursiradiatus* beds, Smithian.

Derivation of name. – From the Latin word: ampliatus meaning enlarged.

Occurrence. – Jin30; *Flemingites rursiradiatus* beds.

Description. – Involute, very small platycone with a circular venter, rounded ventral shoulders and slightly convex flanks. Whorl width slightly thicker than *P. pygmaeus*. Umbilicus with moderately high wall and rounded shoulders. Ornamentation consists only of prorsiradiate and projected concave constrictions.

Procurvoceratites subtabulatus n. sp.

Pl. 44: 7; Table 1

Diagnosis. – *Procurvoceratites* with forward projected, concave constrictions and a subtabulate venter.

Holotype. – PIMUZ 26199, Loc. Jin30, Jinya, *Flemingites rursiradiatus* beds, Smithian.

Derivation of name. – Species name refers to its subtabulate venter.

Occurrence. – Jin30; *Flemingites rursiradiatus* beds.

Description. – Involute, compressed platycone with a subtabulate venter, abruptly rounded ventral shoulders and nearly parallel flanks. Whorl height greater than *P. pygmaeus*. Umbilical depth moderately shallow with perpendicular wall and rounded shoulders. Ornamentation consists of prorsiradiate, projected, concave constrictions that form small crenulations as they cross the subtabulate venter. Suture line unknown.

Discussion. – This species is closely linked to *P. pygmaeus* and may represent a laterally compressed variant. However, large enough samples documenting such a large intraspecific variation are not available.

Genus gen. indet. A

Pl. 45: 1

Occurrence. – Jin12; *Owenites koeneni* beds.

Description. – Single specimen consisting only of a portion of the body chamber. Very evolute, serpenticone with a circular venter, rounded ventral and umbilical shoulders and gently convex flanks. Ornamentation consists of deep, distant (five to six on half volution), concave constrictions that form a somewhat corrugated surface on inner mold. Constrictions tend to be less pronounced near aperture. Inner whorls and suture line unknown.

Genus gen. indet. B

Pl. 45: 4

Occurrence. – Yu22; *Anasibirites multiformis* beds.

Description. – Single specimen only represented by a portion of body chamber. Evolute, compressed platycone with a narrowly rounded venter, rounded ventral and umbilical shoulders, and slightly convex flanks. Ornamentation consists of only large folds. Inner whorls and suture line unknown.

Genus gen. indet. C

Pl. 45: 2

Occurrence. – Yu22; *Anasibirites multiformis* beds.

Description. – Small, evolute, compressed platycone with a narrowly rounded venter, rounded ventral and umbilical shoulders, and weakly convex flanks with maximum curvature near venter. Our single specimen is fragmentary, but umbilicus appears relatively shallow with oblique wall. Ornamentation consists of distinctive, straight, radial ribs. Suture line unknown.

Genus gen. indet. D

Pl. 45: 3

Occurrence. – Yu22; *Anasibirites multiformis* beds.

Description. – Genus only represented by part of body chamber. Evolute serpenticone with a low-rounded venter, rounded ventral and umbilical shoulders, and weakly convex flanks. Umbilicus apparently moderately deep with perpendicular wall. Ornamentation consists of weak, distant, and slightly projected plications, more pronounced near venter. Suture line unknown.

Order Phylloceratitida Zittel, 1884

Superfamily Ussuritaceae Hyatt, 1900

Family Palaeophyllitidae Popov, 1958

?Palaeophyllitidae gen. indet.

Pl. 7: 7a–d; Table 1

Occurrence. – Jin47; Yu7; *Owenites koeneni* beds.

Description. – Evolute, compressed shell with an ovoid whorl section, a circular venter, rounded ventral shoulders and weakly convex, convergent flanks. Umbilicus moderately deep with perpendicular wall and rounded shoulders. Ornamentation consists of more or less dense, almost straight plications, as well as dense radial lirae, which are easily visible at all developmental stages. Plications appear denser and more regular on inner whorls. Suture line displays phylloid saddles and broad lateral lobe. Lobes are well indented.

Discussion. – The combination of shell shape and radial lirae displayed by this specimen is somewhat reminiscent of some Spathian Palaeophyllitidae. In addition, the suture line exhibits phylloid saddles and a structural scheme close to those of Spathian Palaeophyllitidae. Upon evaluation of these diagnostic characteristics, we have assigned it, with some reservation, to Palaeophyllitidae, but additional specimens must be collected in order to confirm this assignment. However, this new Chinese specimen does suggest that the origin of Palaeophyllitidae could be older than previously thought (upper Smithian *contra* Spathian).

Acknowledgements. – Kuang Guodun (Nanning) is gratefully acknowledged for his enthusiasm and invaluable assistance in the field. Thomas Galfetti and Nicolas Goudemand (PIMUZ) are also thanked for their help in the field. We are grateful to T. Galfetti for his work on Figures 1–5 and 7–12. Jim Jenks (Salt Lake City) is thanked for the loan of comparative material and for improving the English text. Claude Monnet (PIMUZ) is thanked for the use of his statistical analysis software. Gilles Escarguel (Lyon) and Thomas Brühwiler (PIMUZ) provide both helpful comments on earlier versions of the manuscript. F. Stiller (Nanjing) kindly opened the doors of Chao's collection. Technical support for photography and preparation was provided by Heinz Lanz, Rosi Roth, Markus Hebeisen, Julia Huber and Leonie Pauli (PIMUZ). Finally, two anonymous reviewers, David A.T. Harper and Svend Stouge provided constructive suggestions and editorial assistance that helped us to improve the manuscript. This work is a contribution to the Swiss National Science Foundation project 200020-113554 (H.B.). A.B. has also benefited from a region Rhône-Alpes Eurodoc grant. Most of the publication costs were covered by the Swiss National Science Foundation. The Zürcher Universitätsverein is also thanked for its contribution to the publication cost. Contribution UMR5125-07.053.

References

Arthaber, G.V. 1911: Die Trias von Albanien. *Beiträge zur Paläontologie und Geologie Österreich-Ungarns und des Orients 24*, 169–276.

Bando, Y. 1964: The Triassic stratigraphy and ammonite fauna of Japan. *The Science Reports of the Tohoku University, Sendai, Japan - Second Series (Geology) 36*, 1–137.

Bando, Y. 1981: Lower Triassic ammonoids from Guryul Ravine and the Spur three kilometres north of Burus. *In* Nakazawa, K. & H.R. Kapoor (eds): *The Upper Permian and Lower Triassic Faunas of Kashmir. Palaeontologia Indica, New Series 46*, 137–171.

Bengtson. P. 1988: Open nomenclature. *Palaeontology 31*, 223–227.

Boulin, J. 1991: Structures in Southwest Asia and evolution of the Eastern Tethys. *Tectonophysics 196*, 211–268.

Brayard, A., Héran, M.-A., Costeur, L. & Escarguel, G. 2004: Triassic and Cenozoic palaeobiogeography: two case studies in quantitative modelling using IDL. *Palaeontologia Electronica 7*, 22.

Brayard, A., Bucher, H., Brühwiler, T., Galfetti, T., Goudemand, N., Guodun, K., Escarguel, G. & Jenks, J. 2007a: *Proharpoceras* Chao: a new ammonoid lineage surviving the end-Permian mass extinction. *Lethaia 40*, 175–181.

Brayard, A., Bucher, H., Escarguel, G., Fluteau, F., Bourquin, S. & Galfetti, T. 2006: The Early Triassic ammonoid recovery: paleoclimatic significance of diversity gradients. *Palaeogeography, Palaeoclimatology, Palaeoecology 239*, 374–395.

Brayard, A., Escarguel, G. & Bucher, H. 2007b: The biogeography of Early Triassic ammonoid faunas: clusters, gradients, and networks. *Geobios 40*, 749–765.

Burij, I.V. & Zharnikova, N.K. 1968: On findings of *Anasibirites* – fauna in South Primorye and its stratigraphic position. *In*

Burij, I.V., Zakharov, Y.D., Zharnikova, N.K. & Nevolin, L.A. (eds): *Sedimentary and Volcanogenic - Sedimentary Formations*, (Ustinovskyi Y.B., ed.). (in Russian) 79–81. DVFAN Nauka, Vladivostok, Russia.

Chao, K. 1950: Some new ammonite genera of Lower Triassic from western Kwangsi. *Palaeontological Novitates 5*, 1–11.

Chao, K. 1959: Lower Triassic ammonoids from Western Kwangsi, China. *Palaeontologia Sinica, New Series B 9*, 355.

Collignon, M. 1933–1934: Paléontologie de Madagascar XX – Les céphalopodes du Trias inférieur. *Annales de Paléontologie 12-13*, 151–162 & 1–43.

Collignon, M. 1973: Ammonites du Trias inférieur et moyen d'Afghanistan. *Annales de Paléontologie 59*, 127–163.

Dagys, A.S., Bucher, H. & Weitschat, W. 1999: Intraspecific variation of *Parasibirites kolymensis* Bychkov (Ammonoidea) from the Lower Triassic (Spathian) of Arctic Asia. *Mitteilungen aus dem Geologisch - Paläontologischen Institut - Universität Hamburg 83*, 163–178.

Dagys, A.S. & Ermakova, S.P. 1988: Boreal Late Olenekian ammonoids. *Transactions of the Academy of Sciences of the USSR 714*, 136 (in Russian).

Dagys, A.S. & Ermakova, S.P. 1990: *Early Olenekian Ammonoids of Siberia*, 112 pp. (in Russian) Nauka, Moskow.

Dagys, A.S. & Konstantinov, A.G. 1984: The genus *Dieneroceras* in the Lower Triassic. *Institute of Geology and Geophysics of the Siberian Division of the USSR, Academy of Sciences 600*, 27–40.

Dagys, A.S. & Weitschat, W. 1993: Intraspecific variation in Boreal Triassic ammonoids. *Geobios, MS 15*, 107–109.

Diener, C. 1895: Triasdische Cephalopodenfaunen der ostsibirischen küstenprovinz. *Mémoires du Comité Géologique 14*, 1–59.

Diener, C. 1897: Part I: the cephalopoda of the lower trias. *Palaeontologia Indica 15. Himalayan fossils 2*, 1–181.

Diener, C. 1915: *Fossilium Catalogus I, Animalia*, Part 8, Cephalopoda Triadica, 369 pp. W. Junk, Berlin.

Enkin, R.J., Yang, Z., Chen, Y. & Courtillot, V. 1992: Paleomagnetic constraints on the geodynamic history of the major blocks of China from the Permian to the Present. *Journal of Geophysical Research 97*, 13953–13989.

Ermakova, S.P. 2002: *Zonal standard of the Boreal Lower Triassic*, 109 pp. (in Russian) Nauka, Moscow.

Fan, P.-F. 1978: Outline of the tectonic evolution of southwestern China. *Tectonophysics 45*, 261–267.

Frebold, H. 1930: Die altersstellung des fischhorizontes, des grippianiveaus und des unteren saurierhorizontes in Spitzbergen. *Skrifter om Svalbard og Ishavet 28*, 1–36.

Frech, F. 1902: Die Dyas: Lethaea geognostica. Theil 1. *Lethaea Palaeozoica 2*, 579–788.

Galfetti, T., Bucher, H., Brayard, A., Hochuli, P.A., Weissert, H., Guodun, K., Atudorei, V. & Guex, J. 2007a: Late Early Triassic climate change: insights from carbonate carbon isotopes, sedimentary evolution and ammonoid paleobiogeography. *Palaeogeography, Palaeoclimatology, Palaeoecology 243*, 394–411.

Galfetti, T., Bucher, H., Ovtcharova, M., Schaltegger, U., Brayard, A., Brühwiler, T., Goudemand, N., Weissert, H., Hochuli, P.A., Cordey, F. & Guodun, K. 2007b: Timing of the Early Triassic carbon cycle perturbations inferred from new U–Pb ages and ammonoid biochronozones. *Earth and Planetary Science Letters 258*, 593–604.

Galfetti, T., Hochuli, P.A., Brayard, A., Bucher, H., Weissert, H. & Vigran, J.O. 2007c: Smithian-Spathian boundary event: Evidence for global climatic change in the wake of the end-Permian biotic crisis. *Geology 35*, 291–294.

Gilder, S.A., Coe, R.S., Wu, H., Guodun, K., Zhao, X. & Wu, Q. 1995: Triassic paleomagnetic data from South China and their bearing on the tectonic evolution of the western circum-Pacific region. *Earth and Planetary Sciences Letters 131*, 269–287.

Guex, J. 1978: Le Trias inférieur des Salt Ranges (Pakistan): problèmes biochronologiques. *Eclogae Geologicae Helvetiae 71*, 105–141.

Hada, S. 1966: Discovery of Early Triassic ammonoids from Gua Musang, Kelantan, Malaya. *Journal of Geosciences, Osaka City University 9*, 111–113.

Hammer, Ø. & Bucher, H. 2005: Buckman's first law of covariation – a case of proportionality. *Lethaia 38*, 67–72.

Hsü, T.-Y. 1940: Some Triassic sections of Kueichow. *Bulletin of the Geological Society of China 20*, 161–172.

Hsü, T.-Y. 1943: The Triassic formations of Kueichou. *Bulletin of the Geological Society of China 23*, 121–128.

Hyatt, A. 1900: Cephalopoda. *In* von Zittel, K.A. (ed.): *Textbook of Paleontology*, 502–604. McMillan, London.

Hyatt, A. & Smith, J.P. 1905: The Triassic cephalopod genera of America. *USGS Professional Paper 40*, 1–394.

Jeannet, A. 1959: Ammonites permiennes et faunes triasiques de l'Himalaya central (expédition suisse Arn. Heim et A. Gansser, 1936). *Palaeontologia Indica 34*, 1–189.

Keyserling, A.V. 1845: Beschreibung einiger von Dr. A. Th. v. Middendorff mitgebrachten Ceratiten des arctischen Sibiriens. *Bulletin de l'Académie Impériale des Sciences de St-Pétersbourg 5*, 161–174.

Kiparisova, L.D. 1947: *Atlas of the Guide Forms of the Fossil Faunas of the USSR 7, Triassic.* (in Russian) All Union Scientific Geological Research Institute (VSEGEI), Leningrad (now St. Petersburg), Russia.

Kiparisova, L.D. 1961: Palaeontological fundamentals for the stratigraphy of Triassic deposits of the Primorye region, I, Cephalopod Mollusca. *Transactions of the All Union Scientific Geological Research Institute (VSEGEI) new series 48*, 1–278 (in Russian) Leningrad (now St. Petersburg), Russia.

Klug, C., Brühwiler, T., Korn, D., Schweigert, G., Brayard, A. & Tilsley, J. 2007: Ammonoid shell structures of primary organic composition. *Palaeontology 50*, 1463–1478.

de Koninck, L.G. 1863: Descriptions of some fossils from India, discovered by Dr. A. Fleming, of Edinburg. *Quaterly Journal of the Geological Society of London 19*, 1–19.

Korchinskaya, M.V. 1982: Explanatory note on the biostratigraphic scheme of the Mesozoic (Trias) of Spitsbergen. *USSR Ministry of Geology, PGO Sevmorgeologia* 40–99.

Krafft, A.V. & Diener, C. 1909: Lower Triassic cephalopoda from Spiti, Malla Johar, and Byans. *Palaeontologia Indica 6*, 1–186.

Krystyn, L., Bhargava, O.N. & Richoz, S. 2007: A candidate GSSP for the base of the Olenekian Stage: Mud at Pin Valley; district Lahul and Spiti, Himachal Pradesh (Western Himalaya), India. *Albertiana 35*, 5–20.

Kuenzi, W.D. 1965: Early Triassic (Scythian) ammonoids from northeastern Washington. *Journal of Paleontology, 39*, 365–378.

Kummel, B. 1952: A classification of the Triassic ammonoids. *Journal of Paleontology 26*, 847–853.

Kummel, B. 1957: Systematic descriptions, L138–L139, L142–L143. *In* Arkell, W.J., Furnish, W.M., Kummel, B., Miller, A.K., Moore, R.C., Schindewolf, O., Sylvester-Bradley, P.C. & Wright, C.W. (eds): *Cephalopoda Ammonoidea. Treatise on Invertebrate Paleontology* (Moore, R.C. ed.): Part L, Mollusca 4. 490 pp. Geological Society of America, Boulder, Colorado, and University of Kansas Press, Lawrence, Kansas.

Kummel, B. 1961: The Spitsbergen arctoceratids. *Bulletin of the Museum of Comparative Zoology 123*, 499–532.

Kummel, B. & Erben, H.K. 1968: Lower and Middle Triassic cephalopods from Afghanistan. *Palaeontographica 129*, 95–148.

Kummel, B. & Sakagami, S. 1960: Mid-Scythian ammonites from Iwai formation, Japan. *Breviora, Museum of Comparative Zoology 126*, 1–13.

Kummel, B. & Steele, G. 1962: Ammonites from the *Meekoceras gracilitatus* zone at Crittenden Spring, Elko County, Nevada. *Journal of Paleontology 36*, 638–703.

Lehrmann, D.J., Payne, J.L., Felix, S.V., Dillett, P.M., Hongmei, W., Youyi, Y. & Jiayong, W. 2003: Permian–Triassic boundary sections from shallow-marine carbonate platforms of the Nanpanjiang basin, South China: implications for oceanic conditions associated with the End-Permian extinction and its aftermath. *Palaios 18*, 138–152.

Lilliefors, H.W. 1967: On the Kolmogorov–Smirnov test for normality with mean and variance unknown. *American Statistical Association Journal 62*, 399–402.

Lindstrom, G. 1865: Om Trias och Juraforsteningar fran Spetsbergen. *Svenska Vetenskap-Akademien Handlingar 6*, 1–20.

Mathews, A.A.L. 1929: The Lower Triassic cephalopod fauna of the Fort Douglas area, Utah. *Walker Museum Memoirs 1*, 1–46.

Matthews, S.C. 1973: Notes on open nomenclature and synonymy lists. *Palaeontology 16*, 713–719.

Mojsisovics, E. 1886: Arktische triasfaunen. *Mémoires de l'Académie Impériale des Sciences de St-Pétersbourg, VIIe série 33*, 1–159.

Monnet, C. & Bucher, H. 2005: New Middle and Late Anisian (Middle Triassic) ammonoid faunas from northwestern Nevada (USA): taxonomy and biochronology. *Fossils and Strata 52*, 121.

Nichols, K.M. & Silberling, N.J. 1979: Early Triassic (Smithian) ammonites of Paleoequatorial affinity from the Chulitna terrane, South-central Alaska. *Geological Survey Professional Paper 1121-B*, B1–B5.

Noetling, F. 1905: Die asiatische Trias. *In* Frech, F. (ed.): *Lethaea geognostica*, 107–221. Verlag der E. Schweizerbart'schen Verlagsbuchhandlung (E. Nägele), Stuttgart, Germany.

Popov, Y. 1961: Triassic ammonoids of northeastern USSR. *Transactions, Scientific Research Institute for the Geology of the Arctic (NIIGA) 79*, 1–179 (in Russian).

Popov, Y. 1962: Some Early Triassic ammonoids of northern Caucasus. *Transactions of the Academy of Sciences of the USSR 127*, 176–184 (in Russian).

Renz, C. & Renz, O. 1948: Eine untertriadische Ammonitenfauna von der griechischen Insel Chios. *Schweizerische Palaeontologische Abhandlungen 66*, 3–98.

Ruzhencev, V.E. 1959: Classification of the superfamily Otocerataceae. *Paleontological Journal 2*, 56–67.

Sakagami, S. 1955: Lower Triassic ammonites from Iwai, Orguno-Mura, Nishitamagun, Kwanto massif, Japan. *Science Reports Tokyo Kyoiku Diagaku, series C 30*, 131–140.

Shevyrev, A.A. 1968: Triassic ammonoidea from the southern part of the USSR. *Transactions of the Palaeontological Institute 119*, 1–272. Nauka, Moscow (in Russian).

Shevyrev, A.A. 1990: Ammonoids and chronostratigraphy of the Triassic. *Trudy Paleontologiceskogo Instituta (Akademija Nauk SSR) 241*, 1–179 (in Russian).

Shevyrev, A.A. 1995: Triassic ammonites of northwestern Caucasus *Trudy Paleontologiceskogo Instituta (Akademija Nauk SSR) 264*, 1–174 (in Russian).

Shevyrev, A.A. 2001: Ammonite zonation and interregional correlation of the Induan stage. *Stratigraphy and Geological Correlation 9*, 473–482.

Shevyrev, A.A. 2006: Triassic biochronology: state of the art and main problems. *Stratigraphy and Geological Correlation 14*, 629–641.

Smith, J.P. 1904: The comparative stratigraphy of the marine Trias of Western America. *Proceedings of the California Academy of Sciences, series 3, 1*, 323–430.

Smith, J.P. 1914: Middle Triassic marine invertebrate faunas of North America. *USGS Professional Paper 83*, 1–254.

Smith, J.P. 1927: Upper Triassic marine invertebrate faunas of North America. *USGS Professional Paper 141*, 1–262.

Smith, J.P. 1932: Lower Triassic ammonoids of North America. *USGS Professional Paper 167*, 1–199.

Spath, L.F. 1930: The Eotriassic invertebrate fauna of east Greenland. *Saertryk af Meddelelser om Grønland 83*, 1–90.

Spath, L.F. 1934: *Part 4: the Ammonoidea of the Trias, Catalogue of the Fossil Cephalopoda in the British Museum (Natural History)*, 521 pp. The Trustees of the British Museum, London.

Tong, J. & Yin, H. 2002: The Lower Triassic of South China. *Journal of Asian Earth Sciences 20*, 803–815.

Tong, J., Yin, H., Zhang, J. & Zhao, L. 2001: Proposed new Lower Triassic stages in South China. *Science in China (Series D) 44*, 961–967.

Tong, J., Zakharov, Y.D. & Wu, S. 2004: Early Triassic ammonoid succession in Chaohu, Anhui province. *Acta Palaeontologica Sinica 43*, 192–204.

Tong, J., Zakharov, Y.D. & Yu, J. 2006: Some additional data to the Lower Triassic of the West Pingdingshan section in Chaohu, Anhui Province, China. *Albertiana 34*, 52–58.

Tozer, E.T. 1967: A standard for Triassic time. *Geologic Survey of Canada Bulletin 156*, 141 pp.

Tozer, E.T. 1981: Triassic ammonoidea: classification, evolution and relationship with Permian and Jurassic forms. *In* House, M.R. & Senior, J.R. (eds): *The ammonoidea*, 65–100. The Systematics Association, London.

Tozer, E.T. 1994: Canadian Triassic ammonoid faunas. *Geologic Survey of Canada Bulletin 467*, 663.

Vu Khuc 1984: *Triassic Ammonoids in Vietnam*, 134 pp. Geoinform and Geodata Institute, Hanoi, Vietnam.

Waagen, W. 1895: Salt-range fossils. Vol 2: Fossils from the ceratite formation. *Palaeontologia Indica 13*, 1–323.

Wang, Y.-G., Chen, C., He, G.-X. & Chen, J. 1981: An outline of the marine Triassic in China. *International Union of Geological Sciences 7*, 1–22.

Wanner, J. 1911: Triascephalopoden von Timor und Rotti. *Neues Jahrbuch für Mineralogie, Geologie und Paläontologie 32*, 177–196.

Waterhouse, J.B. 1996a: The Early and Middle Triassic ammonoid succession of the Himalayas in Western and Central Nepal. Part 2. Systematics studies of the early Middle Scythian. *Palaeontographica A241*, 27–100.

Waterhouse, J.B. 1996b: The Early and Middle Triassic ammonoid succession of the Himalayas in Western and Central Nepal. Part 3. Late Middle Scythian ammonoids. *Palaeontographica A241*, 101–167.

Weitschat, W. & Lehmann, U. 1978: Biostratigraphy of the uppermost part of the Smithian stage (Lower Triassic) at the Botneheia, W-Spitsbergen. *Mitteilungen aus dem Geologisch – Paläontologischen Institut, Universität Hamburg 48*, 85–100.

Welter, O.A. 1922: *Die Ammoniten der Unteren Trias von Timor*, 160 pp. E. Schweizerbart'sche Verlagsbuchhandlung (Erwin Nägele), Stuttgart, Germany.

Westermann, G.E.G. 1966: Covariation and taxonomy of the Jurassic ammonite *Sonninia adicra* (Waagen). *Neues Jahrbuch für Geologie und Paläontologie, Abhandlungen 124*, 289–312.

White, C.A. 1879: Paleontological papers no. 9: fossils from the Jura-Trias of south-eastern Idaho. *U.S. Geological and Geographical Survey of the Territories, Bulletin 5*, 105–117.

White, C.A. 1880: Contributions to invertebrate paleontology, no. 5: Triassic fossils of south-eastern Idaho. *U.S. Geological Survey of the Territories, 12th Annual Report 1*, 105–118.

Yin, H., Sweet, W.C., Glenister, B.F., Kotlyar, G., Kozur, H.W., Newell, N.D., Sheng, J., Yang, Z. & Zakharov, Y.D. 1996: Recommendation of the Meishan section as Global Stratotype Section and Point for basal boundary of Triassic system. *Newsletter in Stratigraphy 34*, 81–108.

Yin, H., Zhang, K., Tong, J., Yang, Z. & Wu, S. 2001: The global stratotype section and point (GSSP) of the Permian–Triassic boundary. *Episodes 24*, 102–114.

Zakharov, Y.D. 1968: *Biostratigraphiya i Amonoidei Nizhnego Triasa Yuzhnogo Primorya (Lower Triassic biostratigraphy and Ammonoids of South Primorye)*, 175 pp. Nauka, Moskva (in Russian).

Zakharov, Y.D. 1978: *Lower Triassic Ammonoids of East USSR*, 224 pp. Nauka, Moskva (in Russian).

Plates 1–45

Plate 1
(All figures natural size)

1a–c: *Kashmirites guangxiense* **n. sp. PIMUZ 25800.**
       Loc. Jin4, Jinya, *Flemingites rursiradiatus* beds, Smithian.

2a–c: *Kashmirites guangxiense* **n. sp. PIMUZ 25801.**
       Loc. Jin4, Jinya, *Flemingites rursiradiatus* beds, Smithian.

3a–b: *Kashmirites guangxiense* **n. sp. PIMUZ 25802.**
       Loc. Jin4, Jinya, *Flemingites rursiradiatus* beds, Smithian.

4a–c: *Kashmirites guangxiense* **n. sp. PIMUZ 25803. Robust variant.**
       Loc. FSB, Jinya, *Flemingites rursiradiatus* beds, Smithian.

5a–d: *Kashmirites guangxiense* **n. sp. PIMUZ 25804.**
       Loc. Jin4, Jinya, *Flemingites rursiradiatus* beds, Smithian.
       a–c) Lateral, ventral and apertural views
       d) Suture line. Scale bar = 5 mm; H = 7 mm.

6a–c: *Kashmirites guangxiense* **n. sp. PIMUZ 25805.**
       Loc. Jin4, Jinya, *Flemingites rursiradiatus* beds, Smithian.

7a–c: *Kashmirites guangxiense* **n. sp. PIMUZ 25806.**
       Loc. Jin4, Jinya, *Flemingites rursiradiatus* beds, Smithian.

8a–b: *Kashmirites guangxiense* **n. sp. PIMUZ 25807.**
       Loc. Jin30, Jinya, *Flemingites rursiradiatus* beds, Smithian.

9a–d: *Kashmirites guangxiense* **n. sp. PIMUZ 25808. Holotype.**
       Loc. WSB, Waili, *Flemingites rursiradiatus* beds, Smithian.

10a–c: *Kashmirites guangxiense* **n. sp. PIMUZ 25809. Paratype.**
       Loc. WSB, Waili, *Flemingites rursiradiatus* beds, Smithian.

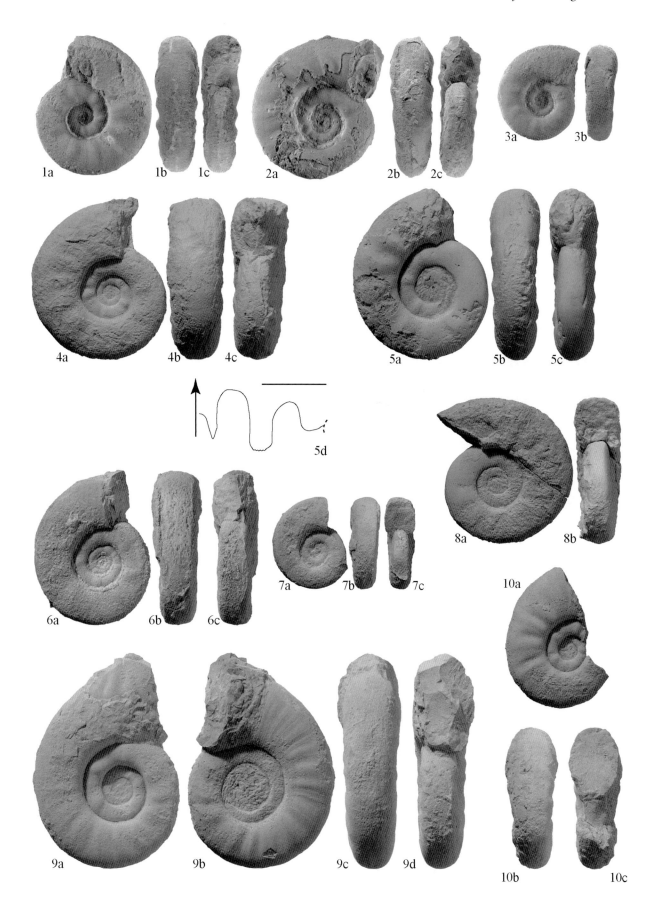

Plate 2
(All figures natural size)

1a–e: *Kashmirites kapila* **(Diener, 1897). PIMUZ 25810. Holotype.**
Loc. Jin64, Waili Cave, *Kashmirites kapila* beds, Smithian.
a–d) Lateral, ventral and apertural views.
e) Suture line. Scale bar = 5 mm; H = 8 mm.

2a–c: *Kashmirites kapila* **(Diener, 1897). PIMUZ 25811.**
Loc. Jin67, Waili Cave, *Kashmirites kapila* beds, Smithian.

3a–b: *Kashmirites kapila* **(Diener, 1897). PIMUZ 25812. Robust variant.**
Loc. Jin67, Waili Cave, *Kashmirites kapila* beds, Smithian.

4a–b: *Kashmirites kapila* **(Diener, 1897). PIMUZ 25813. Robust variant.**
Loc. Jin67, Waili Cave, *Kashmirites kapila* beds, Smithian.

5a–c: *Preflorianites* cf. *radians* **Chao, 1959. PIMUZ 25814.**
Loc. Jin15, Jinya, *Flemingites rursiradiatus* beds, Smithian.

6a–c: *Preflorianites* cf. *radians* **Chao, 1959. PIMUZ 25815.**
Loc. 4, Jinya, *Flemingites rursiradiatus* beds, Smithian.

7a–d: *Preflorianites* cf. *radians* **Chao, 1959. PIMUZ 25816.**
Loc. Jin30, Jinya, *Flemingites rursiradiatus* beds, Smithian.

8a–d: *Preflorianites* cf. *radians* **Chao, 1959. PIMUZ 25817.**
Loc. Jin13, Jinya, *Flemingites rursiradiatus* beds, Smithian.

9a–d: *Preflorianites* cf. *radians* **Chao, 1959. PIMUZ 25818.**
Loc. Jin30, Jinya, *Flemingites rursiradiatus* beds, Smithian.

10a–c: *Preflorianites* cf. *radians* **Chao, 1959. PIMUZ 25819.**
Loc. FSB, Jinya, *Flemingites rursiradiatus* beds, Smithian.

11: **Suture line of *Preflorianites* cf. *radians* Chao, 1959. PIMUZ 25820.**
Loc. Jin30, Jinya, *Flemingites rursiradiatus* beds, Smithian.
Scale bar = 5 mm; H = 6 mm.

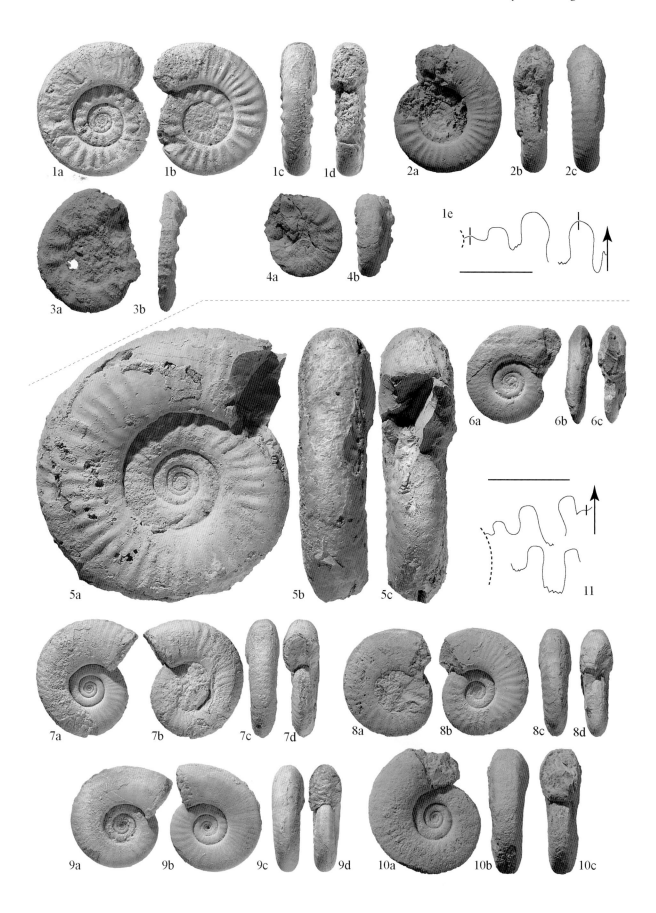

Plate 3
(All figures natural size)

1a–d: ***Pseudoceltites? angustecostatus* (Welter, 1922). PIMUZ 25821. Paratype.**
Loc. T5, Tsoteng, *Owenites koeneni* beds, Smithian.

2a–c: ***Pseudoceltites? angustecostatus* (Welter, 1922). PIMUZ 25822. Holotype.**
Loc. T5, Tsoteng, *Owenites koeneni* beds, Smithian.

3a–c: ***Pseudoceltites? angustecostatus* (Welter, 1922). PIMUZ 25823. Paratype.**
Loc. T5, Tsoteng, *Owenites koeneni* beds, Smithian.

4a–c: ***Pseudoceltites? angustecostatus* (Welter, 1922). PIMUZ 25824. Paratype.**
Loc. T5, Tsoteng, *Owenites koeneni* beds, Smithian.

5a–c: ***Pseudoceltites? angustecostatus* (Welter, 1922). PIMUZ 25825. Paratype.**
Loc. T5, Tsoteng, *Owenites koeneni* beds, Smithian.

6a–d: ***Pseudoceltites? angustecostatus* (Welter, 1922). PIMUZ 25826. Paratype.**
Loc. T5, Tsoteng, *Owenites koeneni* beds, Smithian.
a–c) Lateral, ventral and apertural views.
d) Suture line. Scale bar = 5 mm; H = 5 mm.

7a–b: ***Pseudoceltites? angustecostatus* (Welter, 1922). PIMUZ 25827. Paratype.**
Loc. T5, Tsoteng, *Owenites koeneni* beds, Smithian.

Plate 4
(All figures natural size unless otherwise indicated)

1a–d: *Hanielites elegans* **Welter, 1922. PIMUZ 25828.**
Loc. Jin45, Jinya, *Owenites koeneni* beds, Smithian.

2a–c: *Hanielites elegans* **Welter, 1922. PIMUZ 25829. Scale ×2.**
Loc. Jin45, Jinya, *Owenites koeneni* beds, Smithian.

3a–c: *Hanielites elegans* **Welter, 1922. PIMUZ 25830. Scale ×2.**
Loc. Jin45, Jinya, *Owenites koeneni* beds, Smithian.

4a–d: *Hanielites elegans* **Welter, 1922. PIMUZ 25831. Scale ×2.**
Loc. Jin45, Jinya, *Owenites koeneni* beds, Smithian.

5: **Suture line of** *Hanielites elegans* **Welter, 1922. PIMUZ 25837.**
Loc. Jin45, Jinya, *Owenites koeneni* beds, Smithian.
Scale bar = 2.5 mm; H = 5 mm.

6a–d: *Hanielites gracilus* **n. sp. PIMUZ 25833. Paratype.**
Loc. Jin45, Jinya, *Owenites koeneni* beds, Smithian.
6d: Scale ×4.

7a–d: *Hanielites gracilus* **n. sp. PIMUZ 25834. Holotype.**
Loc. Jin45, Jinya, *Owenites koeneni* beds, Smithian.
a–c) Lateral, ventral and apertural views.
d) Suture line. Scale bar = 2.5 mm; H = 10 mm.

8a–c: *Hanielites gracilus* **n. sp. PIMUZ 25835.**
Loc. Jin10, Jinya, *Flemingites rursiradiatus* beds, Smithian.

9a–c: *Hanielites carinatitabulatus* **Chao, 1959. PIMUZ 25832. Scale ×2.**
Loc. Yu1, Yuping, *Owenites koeneni* beds, Smithian.
a–b) Lateral and ventral views.
c) Suture line. Scale bar = 2.5 mm; H = 6 mm.

10a–d: *Hanielites angulus* **n. sp. PIMUZ 25836. Holotype.**
Loc. Yu1, Yuping, *Owenites koeneni* beds, Smithian.
10d: Scale ×3.

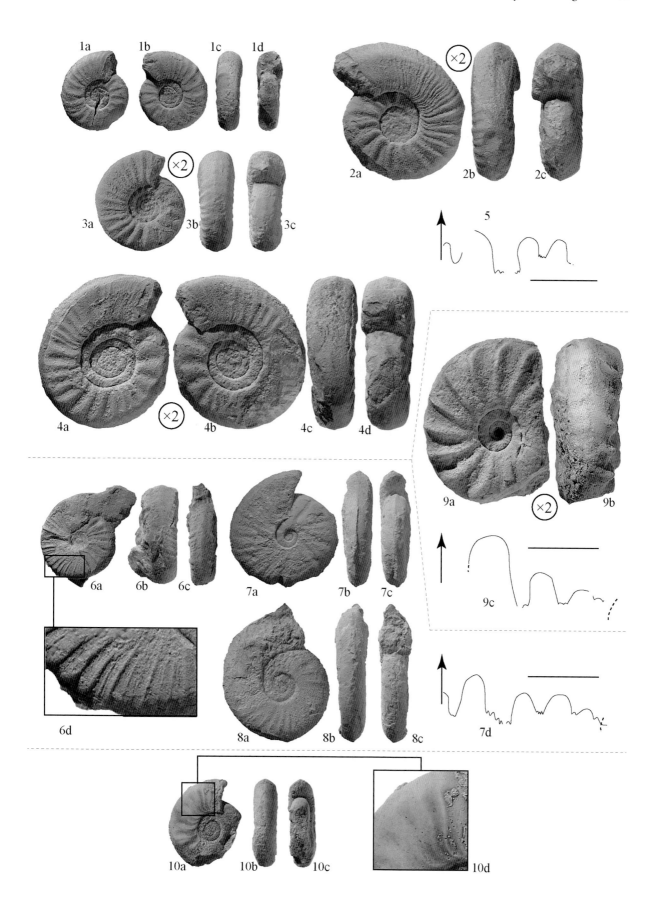

Plate 5
(All figures natural size unless otherwise indicated)

1a–d: *Xenoceltites variocostatus* **n. sp. PIMUZ 25838. Holotype.**
Loc. NW13, Waili, *Anasibirites multiformis* beds, Smithian.

2a–c: *Xenoceltites variocostatus* **n. sp. PIMUZ 25839. Scale ×2. Paratype.**
Loc. NW13, Waili, *Anasibirites multiformis* beds, Smithian.

3a–c: *Xenoceltites variocostatus* **n. sp. PIMUZ 25840. Scale ×2. Paratype.**
Loc. NW13, Waili, *Anasibirites multiformis* beds, Smithian.

4a–c: *Xenoceltites variocostatus* **n. sp. PIMUZ 25841. Paratype.**
Loc. NW13, Waili, *Anasibirites multiformis* beds, Smithian.

5:   *Xenoceltites variocostatus* **n. sp. PIMUZ 25842. Paratype.**
Loc. NW13, Waili, *Anasibirites multiformis* beds, Smithian.

6a–c: *Xenoceltites variocostatus* **n. sp. PIMUZ 25843. Paratype.**
Loc. NW13, Waili, *Anasibirites multiformis* beds, Smithian.

7a–c: *Xenoceltites variocostatus* **n. sp. PIMUZ 25844. Paratype.**
Loc. NW13, Waili, *Anasibirites multiformis* beds, Smithian.

8a–c: *Xenoceltites variocostatus* **n. sp. PIMUZ 25845. Scale ×2. Paratype.**
Loc. NW13, Waili, *Anasibirites multiformis* beds, Smithian.

9a–c: *Xenoceltites variocostatus* **n. sp. PIMUZ 25846. Paratype.**
Loc. NW13, Waili, *Anasibirites multiformis* beds, Smithian.

10a–d: *Xenoceltites variocostatus* **n. sp. PIMUZ 25847. Scale ×2. Paratype.**
Loc. NW13, Waili, *Anasibirites multiformis* beds, Smithian.

11a–c: *Xenoceltites variocostatus* **n. sp. PIMUZ 25848. Paratype.**
Loc. NW13, Waili, *Anasibirites multiformis* beds, Smithian.

12:   *Xenoceltites variocostatus* **n. sp. PIMUZ 25849. Paratype.**
Loc. NW13, Waili, *Anasibirites multiformis* beds, Smithian.

13a–d: *Xenoceltites variocostatus* **n. sp. PIMUZ 25850. Scale ×2. Paratype.**
Loc. NW13, Waili, *Anasibirites multiformis* beds, Smithian.

14a–c: *Xenoceltites variocostatus* **n. sp. PIMUZ 25851. Paratype.**
Loc. NW13, Waili, *Anasibirites multiformis* beds, Smithian.

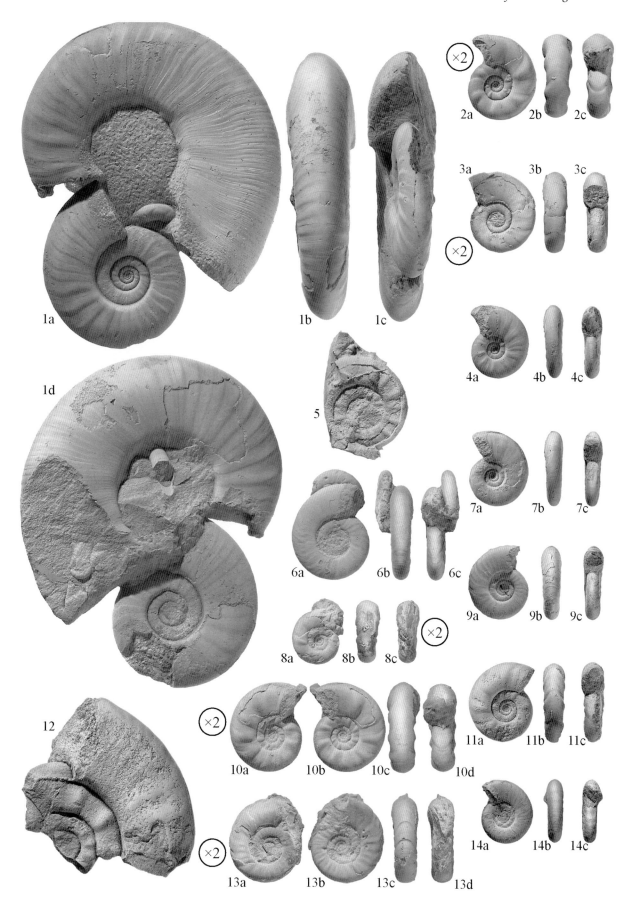

Plate 6
(All figures natural size unless otherwise indicated)

1a–b: *Xenoceltites variocostatus* **n. sp. PIMUZ 25852. Scale ×2. Paratype.**
Loc. NW13, Waili, *Anasibirites multiformis* beds, Smithian.

2a–c: *Xenoceltites variocostatus* **n. sp. PIMUZ 25853. Paratype.**
Loc. NW13, Waili, *Anasibirites multiformis* beds, Smithian.

3a–c: *Xenoceltites variocostatus* **n. sp. PIMUZ 25854. Paratype.**
Loc. NW13, Waili, *Anasibirites multiformis* beds, Smithian.

4a–c: *Xenoceltites variocostatus* **n. sp. PIMUZ 25855.**
Loc. Jin101, Waili, *Anasibirites multiformis* beds, Smithian.

5a–c: *Xenoceltites variocostatus* **n. sp. PIMUZ 25856. Paratype.**
Loc. NW13, Waili, *Anasibirites multiformis* beds, Smithian.

6: **Suture line of *Xenoceltites variocostatus* n. sp., PIMUZ 25857. Paratype.**
Loc. NW13, Waili, *Anasibirites multiformis* beds, Smithian.
Scale bar = 5 mm; D = 12 mm.

7a–c: *Xenoceltites pauciradiatus* **n. sp. PIMUZ 25858. Holotype.**
Loc. Jin33, Jinya, *Anasibirites multiformis* beds, Smithian.

8a–c: *Xenoceltites pauciradiatus* **n. sp. PIMUZ 25859. Paratype.**
Loc. Jin33, Jinya, *Anasibirites multiformis* beds, Smithian.

9a–c: *Xenoceltites pauciradiatus* **n. sp. PIMUZ 25860. Paratype.**
Loc. Jin33, Jinya, *Anasibirites multiformis* beds, Smithian.

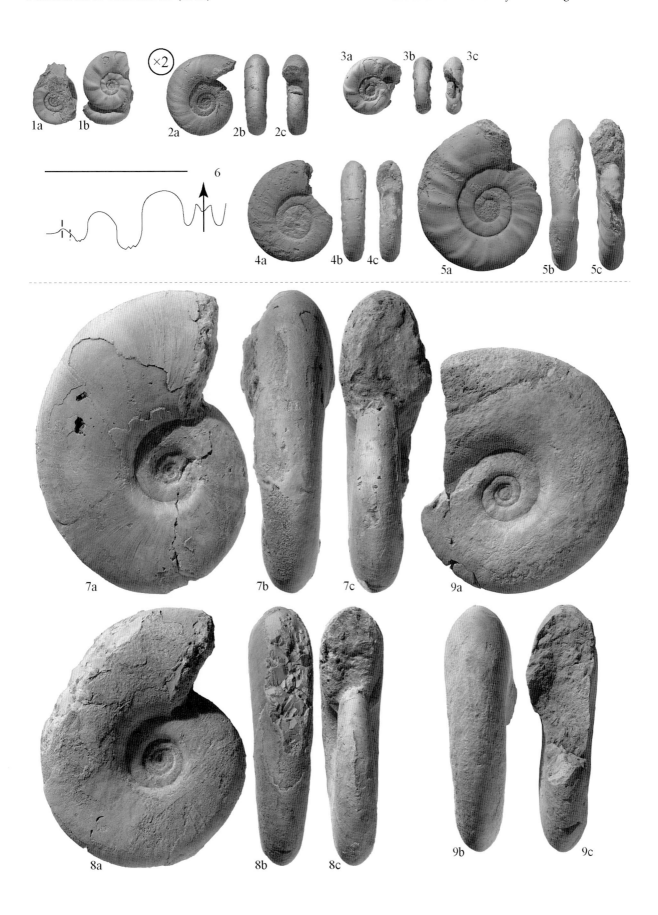

Plate 7
(All figures natural size unless otherwise indicated)

1a–d: *Guangxiceltites admirabilis* **n. gen., n. sp. PIMUZ 25861. Holotype.**
Loc. Jin61, Jinya, *Kashmirites kapila* beds, Smithian.

2a–c: *Guangxiceltites admirabilis* **n. gen., n. sp. PIMUZ 25862. Paratype.**
Loc. Jin61, Jinya, *Kashmirites kapila* beds, Smithian.

3a–c: *Guangxiceltites admirabilis* **n. gen., n. sp. PIMUZ 25863. Scale ×1.5. Paratype.**
Loc. Jin61, Jinya, *Kashmirites kapila* beds, Smithian.

4a–c: *Guangxiceltites admirabilis* **n. gen., n. sp. PIMUZ 25864. Paratype.**
Loc. Jin61, Jinya, *Kashmirites kapila* beds, Smithian.

5a–c: *Guangxiceltites admirabilis* **n. gen., n. sp. PIMUZ 25865. Paratype.**
Loc. Jin61, Jinya, *Kashmirites kapila* beds, Smithian.

6a–c: *Guangxiceltites admirabilis* **n. gen., n. sp. PIMUZ 25866.**
Loc. Jin64, Jinya, *Kashmirites kapila* beds, Smithian.

7a–d: **?Palaeophyllitidae gen. indet. PIMUZ 25867. Holotype.**
Loc. Jin47, Jinya, *Owenites koeneni* beds, Smithian.
a–c) Lateral, ventral and apertural views. Scale ×1.25
d) Suture line. Scale bar = 5 mm; H = 13 mm.

8a–d: **Xenoceltitidae gen. indet. PIMUZ 25868.**
Loc. Yu22, Yuping, *Kashmirites kapila* beds, Smithian.
a–c) Lateral and ventral views. Scale ×0.75.
d) Suture line. Scale bar = 5 mm; H = 16 mm.

9a–d: *Weitschaticeras concavum* **n. gen., n. sp. PIMUZ 25869. Holotype.**
Loc. Jin27, Jinya, *Owenites koeneni* beds, Smithian.
a–c) Lateral, ventral and apertural views.
d) Suture line. Scale bar = 5 mm; H = 9 mm.

Plate 8
(All figures natural size unless otherwise indicated)

1a–c: *Hebeisenites varians* **(Chao, 1959) PIMUZ 25870.**
Loc. Jin29, Jinya, *Flemingites rursiradiatus* beds, Smithian.

2a–c: *Hebeisenites varians* **(Chao, 1959) PIMUZ 25871. Scale ×2.**
Loc. Jin29, Jinya, *Flemingites rursiradiatus* beds, Smithian.

3a–c: *Hebeisenites varians* **(Chao, 1959) PIMUZ 25872.**
Loc. Jin30, Jinya, *Flemingites rursiradiatus* beds, Smithian.

4a–c: *Hebeisenites varians* **(Chao, 1959) PIMUZ 25873.**
Loc. Jin29, Jinya, *Flemingites rursiradiatus* beds, Smithian.

5a–c: *Hebeisenites varians* **(Chao, 1959) PIMUZ 25874.**
Loc. Jin4, Jinya, *Flemingites rursiradiatus* beds, Smithian.

6a–b: *Hebeisenites varians* **(Chao, 1959) PIMUZ 25875.**
Loc. Jin4, Jinya, *Flemingites rursiradiatus* beds, Smithian.

7a–b: *Hebeisenites varians* **(Chao, 1959) PIMUZ 25876.**
Loc. Jin4, Jinya, *Flemingites rursiradiatus* beds, Smithian.

8a–b: *Hebeisenites varians* **(Chao, 1959) PIMUZ 25877.**
Loc. Jin4, Jinya, *Flemingites rursiradiatus* beds, Smithian.

9a–b: *Hebeisenites varians* **(Chao, 1959) PIMUZ 25878.**
Loc. Jin28, Jinya, *Flemingites rursiradiatus* beds, Smithian.

10: **Suture line of** *Hebeisenites varians* **(Chao, 1959), PIMUZ 25879.**
Loc. Jin23, Jinya, *Flemingites rursiradiatus* beds, Smithian.
Scale bar = 5 mm; D = 15 mm.

11: **Suture line of** *Hebeisenites varians* **(Chao, 1959), PIMUZ 25880.**
Loc. Jin30, Jinya, *Flemingites rursiradiatus* beds, Smithian.
Scale bar = 5 mm; D = 14 mm.

12a–e: *Hebeisenites evolutus* **n. gen., n. sp. PIMUZ 25881.**
Loc. Jin28, Jinya, *Flemingites rursiradiatus* beds, Smithian.
a–d) Lateral, ventral and apertural views.
e) Suture line. Scale bar = 5 mm; H = 2.5 mm.

13a–c: *Hebeisenites evolutus* **n. gen., n. sp. PIMUZ 25882.**
Loc. Jin28, Jinya, *Flemingites rursiradiatus* beds, Smithian.

14a–c: *Hebeisenites evolutus* **n. gen., n. sp. PIMUZ 25883.**
Loc. Jin28, Jinya, *Flemingites rursiradiatus* beds, Smithian.

15a–c: *Hebeisenites evolutus* **n. gen., n. sp. PIMUZ 25884.**
Loc. Jin29, Jinya, *Flemingites rursiradiatus* beds, Smithian.

16a–c: *Hebeisenites evolutus* **n. gen., n. sp. PIMUZ 25885.**
Loc. Jin30, Jinya, *Flemingites rursiradiatus* beds, Smithian.

17a–d: *Hebeisenites evolutus* **n. gen., n. sp. PIMUZ 25886. Holotype.**
Loc. Jin10, Jinya, *Flemingites rursiradiatus* beds, Smithian.

18a–c: *Hebeisenites compressus* **n. gen., n. sp. PIMUZ 25887. Scale ×2.**
Loc. Jin23, Jinya, *Flemingites rursiradiatus* beds, Smithian.

19a–d: *Hebeisenites compressus* **n. gen., n. sp. PIMUZ 25888. Scale ×2. Holotype.**
Loc. Jin30, Jinya, *Flemingites rursiradiatus* beds, Smithian.

20a–b: *Hebeisenites compressus* **n. gen., n. sp. PIMUZ 25889. Scale ×2. Paratype.**
Loc. Jin30, Jinya, *Flemingites rursiradiatus* beds, Smithian.

21a–c: *Hebeisenites compressus* **n. gen., n. sp. PIMUZ 25890. Scale ×2. Paratype.**
Loc. Jin30, Jinya, *Flemingites rursiradiatus* beds, Smithian.

22a–c: *Hebeisenites compressus* **n. gen., n. sp. PIMUZ 25891. Scale ×2. Paratype.**
Loc. Jin30, Jinya, *Flemingites rursiradiatus* beds, Smithian.

23a–d: *Hebeisenites compressus* **n. gen., n. sp. PIMUZ 25892. Scale ×2. Paratype.**
Loc. Jin30, Jinya, *Flemingites rursiradiatus* beds, Smithian.

24a–c: *Hebeisenites compressus* **n. gen., n. sp. PIMUZ 25893. Scale ×2. Paratype.**
Loc. Jin30, Jinya, *Flemingites rursiradiatus* beds, Smithian.

25: **Suture line of** *Hebeisenites compressus* **n. gen., n. sp., PIMUZ 26204.**
Loc. Jin23, Jinya, *Flemingites rursiradiatus* beds, Smithian.
Scale bar = 2.5 mm; H = 3 mm.

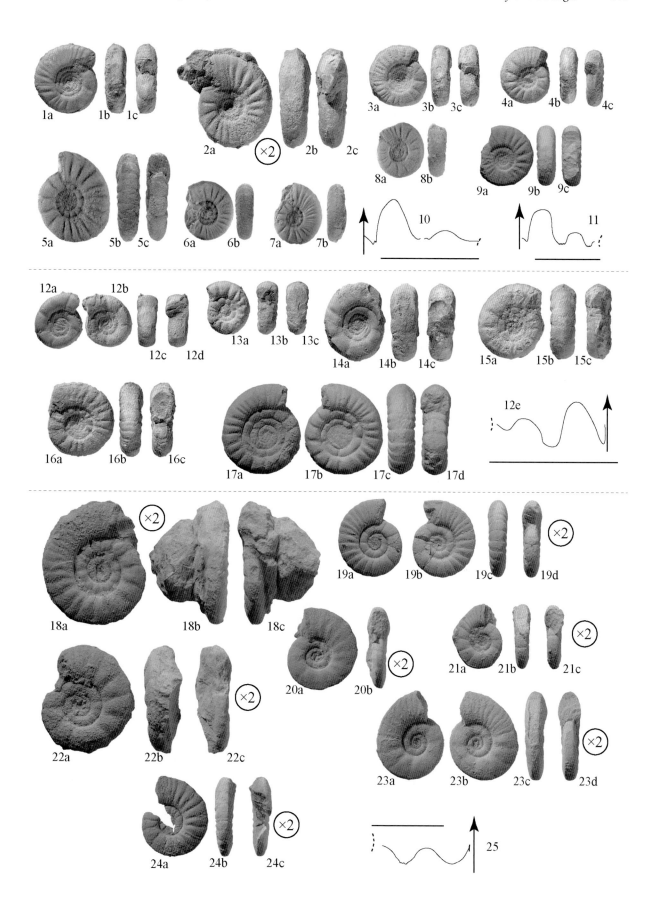

Plate 9
(All figures natural size)

1a–d: *Jinyaceras bellum* **n. gen., n. sp. PIMUZ 25894. Holotype.**
        Loc. Jin28, Jinya, *Flemingites rursiradiatus* beds, Smithian.

2a–c: *Jinyaceras bellum* **n. gen., n. sp. PIMUZ 25895.**
        Loc. Jin30, Jinya, *Flemingites rursiradiatus* beds, Smithian.

3a–c: *Jinyaceras bellum* **n. gen., n. sp. PIMUZ 25896.**
        Loc. Jin4, Jinya, *Flemingites rursiradiatus* beds, Smithian.

4a–c: *Jinyaceras bellum* **n. gen., n. sp. PIMUZ 25897.**
        Loc. Jin30, Jinya, *Flemingites rursiradiatus* beds, Smithian.

5a–c: *Jinyaceras bellum* **n. gen., n. sp. PIMUZ 25898.**
        Loc. Jin30, Jinya, *Flemingites rursiradiatus* beds, Smithian.

6a–c: *Jinyaceras bellum* **n. gen., n. sp. PIMUZ 25899.**
        Loc. Jin30, Jinya, *Flemingites rursiradiatus* beds, Smithian.

7a–c: *Jinyaceras bellum* **n. gen., n. sp. PIMUZ 25900.**
        Loc. T50, Tsoteng, *Flemingites rursiradiatus* beds, Smithian.

8a–c: *Jinyaceras bellum* **n. gen., n. sp. PIMUZ 25901.**
        Loc. Jin30, Jinya, *Flemingites rursiradiatus* beds, Smithian.

9a–c: *Jinyaceras bellum* **n. gen., n. sp. PIMUZ 25902.**
        Loc. Jin30, Jinya, *Flemingites rursiradiatus* beds, Smithian.

10a–c: *Jinyaceras bellum* **n. gen., n. sp. PIMUZ 25903.**
        Loc. Jin4, Jinya, *Flemingites rursiradiatus* beds, Smithian.

11a–c: *Jinyaceras bellum* **n. gen., n. sp. PIMUZ 25904.**
        Loc. Jin4, Jinya, *Flemingites rursiradiatus* beds, Smithian.

12a–c: *Jinyaceras bellum* **n. gen., n. sp. PIMUZ 25905. Paratype.**
        Loc. Jin28, Jinya, *Flemingites rursiradiatus* beds, Smithian.

13a–c: *Jinyaceras bellum* **n. gen., n. sp. PIMUZ 25906. Paratype.**
        Loc. Jin28, Jinya, *Flemingites rursiradiatus* beds, Smithian.

14a–c: *Jinyaceras bellum* **n. gen., n. sp. PIMUZ 25907. Paratype.**
        Loc. Jin28, Jinya, *Flemingites rursiradiatus* beds, Smithian.

15a–c: *Jinyaceras bellum* **n. gen., n. sp. PIMUZ 25908.**
        Loc. Jin30, Jinya, *Flemingites rursiradiatus* beds, Smithian.

16a–c: *Jinyaceras bellum* **n. gen., n. sp. PIMUZ 25909.**
        Loc. Jin30, Jinya, *Flemingites rursiradiatus* beds, Smithian.

17a–c: *Jinyaceras bellum* **n. gen., n. sp. PIMUZ 25910.**
        Loc. Jin30, Jinya, *Flemingites rursiradiatus* beds, Smithian.

18a–c: *Jinyaceras bellum* **n. gen., n. sp. PIMUZ 25911.**
        Loc. Jin30, Jinya, *Flemingites rursiradiatus* beds, Smithian.

19:     **Suture line of** *Jinyaceras bellum* **n. gen., n. sp., PIMUZ 25912.**
        Loc. Jin30, Jinya, *Flemingites rursiradiatus* beds, Smithian.
        Scale bar = 5 mm; H = 4 mm.

20a–c: *Juvenites* **cf.** *kraffti.* **PIMUZ 25913.**
        Loc. Jin23, Jinya, *Flemingites rursiradiatus* beds, Smithian.

21a–c: *Juvenites* **cf.** *kraffti.* **PIMUZ 25914.**
        Loc. Jin30, Jinya, *Flemingites rursiradiatus* beds, Smithian.

22a–c: *Juvenites* **cf.** *kraffti.* **PIMUZ 25915.**
        Loc. Jin30, Jinya, *Flemingites rursiradiatus* beds, Smithian.

23a–c: *Juvenites* **cf.** *kraffti.* **PIMUZ 25916.**
        Loc. Jin4, Jinya, *Flemingites rursiradiatus* beds, Smithian.

24a–d: *Paranorites jenksi* **n. sp. PIMUZ 25917. Holotype.**
        Loc. Jin67, Waili, *Kashmirites kapila* beds, Smithian.
        a–c) Lateral, ventral and apertural views.
        d) Suture line. Scale bar = 5 mm; H = 16 mm.

25a–d: *Paranorites jenksi* **n. sp. PIMUZ 25918.**
        Loc. Jin66, Waili, *Kashmirites kapila* beds, Smithian.

26a–d: *Paranorites jenksi* **n. sp. PIMUZ 25919.**
        Loc. Jin66, Waili, *Kashmirites kapila* beds, Smithian.

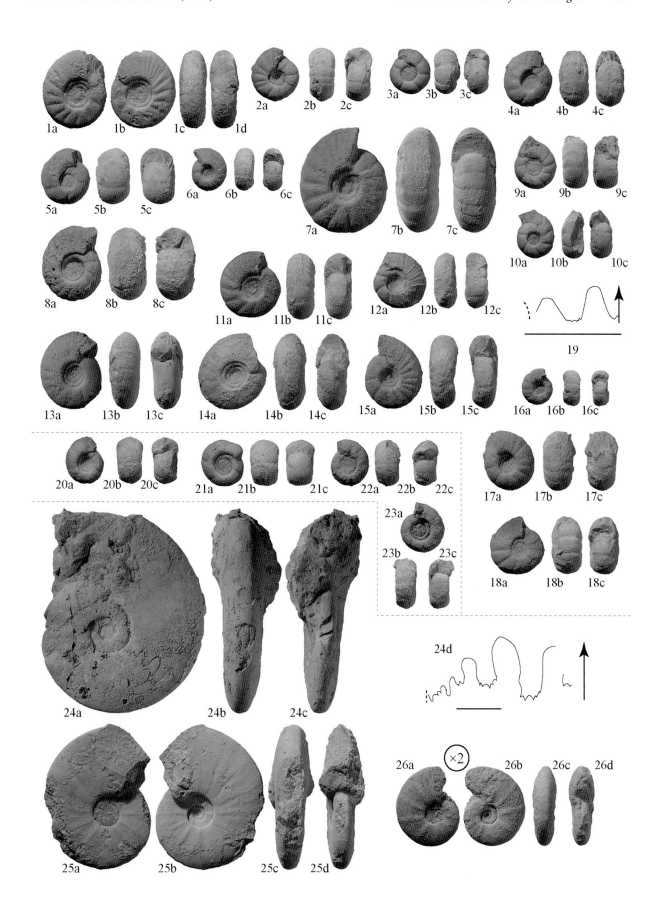

Plate 10
(All figures natural size)

1a–d: *Pseudaspidites muthianus* **(Krafft & Diener, 1909). PIMUZ 25920.**
Loc. Jin28, Jinya, *Flemingites rursiradiatus* beds, Smithian.
a–c) Lateral, ventral and apertural views.
d) Suture line. Scale bar = 5 mm; H = 15 mm. Slightly smoothed.

2a–c: *Pseudaspidites muthianus* **(Krafft & Diener, 1909). PIMUZ 25921.**
Loc. Jin28, Jinya, *Flemingites rursiradiatus* beds, Smithian.

3a–d: *Pseudaspidites muthianus* **(Krafft & Diener, 1909). PIMUZ 25922.**
Loc. Jin28, Jinya, *Flemingites rursiradiatus* beds, Smithian.

4a–c: *Pseudaspidites muthianus* **(Krafft & Diener, 1909). PIMUZ 25923.**
Loc. Jin28, Jinya, *Flemingites rursiradiatus* beds, Smithian.

5a–c: *Pseudaspidites muthianus* **(Krafft & Diener, 1909). PIMUZ 25924.**
Loc. Jin30, Jinya, *Flemingites rursiradiatus* beds, Smithian.

6a–c: *Pseudaspidites muthianus* **(Krafft & Diener, 1909). PIMUZ 25925.**
Loc. Jin30, Jinya, *Flemingites rursiradiatus* beds, Smithian.

7a–e: *Pseudaspidites muthianus* **(Krafft & Diener, 1909). PIMUZ 25928.**
Loc. Jin30, Jinya, *Flemingites rursiradiatus* beds, Smithian.
a–d) Lateral, ventral and apertural views.
e) Suture line. Scale bar = 5 mm; H = 15 mm. Slightly smoothed.

8: **Suture line of *Pseudaspidites muthianus* (Krafft & Diener, 1909). PIMUZ 25927.**
Loc. Jin30, Jinya, *Flemingites rursiradiatus* beds, Smithian.
Scale bar = 5 mm; D = 35 mm. Slightly smoothed.

9: **Suture line of *Pseudaspidites muthianus* (Krafft & Diener, 1909). PIMUZ 25929.**
Loc. Jin30, Jinya, *Flemingites rursiradiatus* beds, Smithian.
Scale bar = 5 mm; H = 35 mm. Slightly smoothed.

10: **Suture line of *Pseudaspidites muthianus* (Krafft & Diener, 1909), variant with umbilical bullae. PIMUZ 25930.**
Loc. Sha1, Shanggan, *Flemingites rursiradiatus* beds, Smithian.
Scale bar = 5 mm; H = 30 mm.

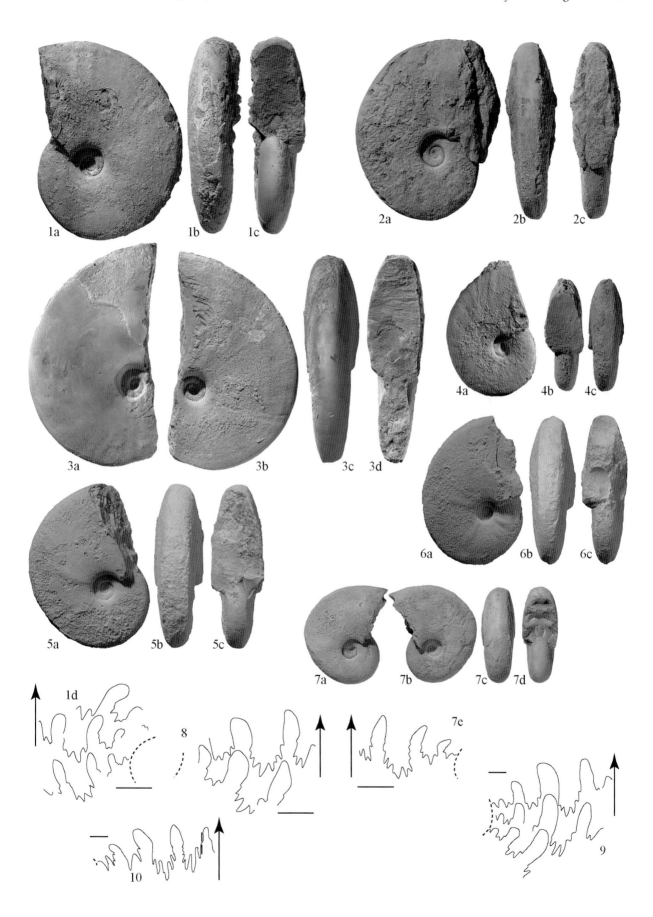

Plate 11
(All figures natural size unless otherwise indicated)

1a–d: *Pseudaspidites muthianus* **(Krafft & Diener, 1909). PIMUZ 25931. Scale ×0.5.**
Loc. Jin4, Jinya, *Flemingites rursiradiatus* beds, Smithian.

2a–d: *Pseudaspidites muthianus* **(Krafft & Diener, 1909). PIMUZ 25930. Robust variant.**
Loc. Sha1, Shanggan, *Flemingites rursiradiatus* beds, Smithian.

3a–c: *Pseudaspidites muthianus* **(Krafft & Diener, 1909). PIMUZ 25932. Robust variant.**
Loc. Jin30, Jinya, *Flemingites rursiradiatus* beds, Smithian.

4a–c: *Pseudaspidites muthianus* **(Krafft & Diener, 1909). PIMUZ 25933. Robust variant.**
Loc. Jin13, Jinya, *Flemingites rursiradiatus* beds, Smithian.

5a–d: Proptychitidae gen. indet. A. PIMUZ 25934.
Loc. Jin30, Jinya, *Flemingites rursiradiatus* beds, Smithian.
a–c) Lateral, ventral and apertural views.
d) Suture line. Scale bar = 5 mm; H = 14 mm.

6a–d: *Pseudaspidites* **sp. indet. PIMUZ 25935.**
Loc. Jin27, Jinya, *Owenites koeneni* beds, Smithian.
a–c) Lateral, ventral and apertural views.
d) Suture line. Scale bar = 5 mm; H = 10 mm.

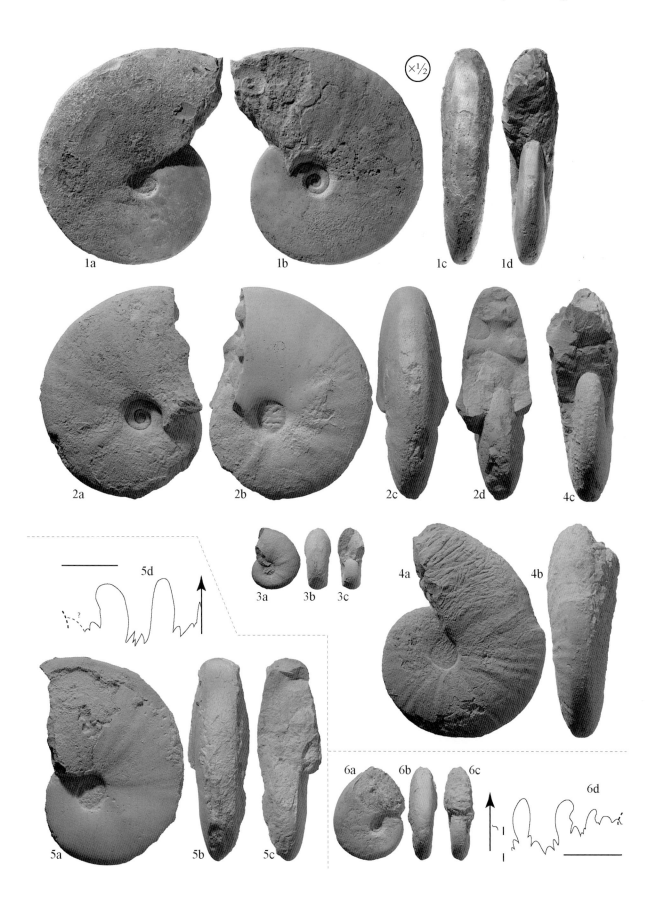

×½

Plate 12
(All figures natural size)

1a–c: *Lingyunites discoides* **Chao, 1950. PIMUZ 25936.**
Loc. Jin30, Jinya, *Flemingites rursiradiatus* beds, Smithian.

2a–c: *Lingyunites discoides* **Chao, 1950. PIMUZ 25937.**
Loc. Jin30, Jinya, *Flemingites rursiradiatus* beds, Smithian.

3a–d: *Lingyunites discoides* **Chao, 1950. PIMUZ 25938.**
Loc. Jin10, Jinya, *Flemingites rursiradiatus* beds, Smithian.

4a–c: *Lingyunites discoides* **Chao, 1950. PIMUZ 25939.**
Loc. Jin30, Jinya, *Flemingites rursiradiatus* beds, Smithian.

5a–c: *Lingyunites discoides* **Chao, 1950. PIMUZ 25940.**
Loc. Jin30, Jinya, *Flemingites rursiradiatus* beds, Smithian.

6a–d: *Lingyunites discoides* **Chao, 1950. PIMUZ 25941.**
Loc. Jin29, Jinya, *Flemingites rursiradiatus* beds, Smithian.
a–c) Lateral, ventral and apertural views.
d) Suture line. Scale bar = 5 mm; H = 9 mm.

7a–c: *Lingyunites discoides* **Chao, 1950. PIMUZ 25942.**
Loc. Jin30, Jinya, *Flemingites rursiradiatus* beds, Smithian.

8a–c: *Lingyunites discoides* **Chao, 1950. PIMUZ 25943.**
Loc. Jin30, Jinya, *Flemingites rursiradiatus* beds, Smithian.

9a–d: *Nanningites tientungense* **n. gen. PIMUZ 25944.**
Loc. Jin29, Jinya, *Flemingites rursiradiatus* beds, Smithian.
a–c) Lateral, ventral and apertural views.
d) Suture line. Scale bar = 5 mm; D = 21 mm.

10a–c: *Nanningites tientungense* **n. gen. PIMUZ 25945.**
Loc. Jin30, Jinya, *Flemingites rursiradiatus* beds, Smithian.

11a–c: *Nanningites tientungense* **n. gen. PIMUZ 25946.**
Loc. Jin30, Jinya, *Flemingites rursiradiatus* beds, Smithian.

12a–c: *Xiaoqiaoceras involutus* **n. gen., n. sp. PIMUZ 25947. Paratype.**
Loc. Jin4, Jinya, *Flemingites rursiradiatus* beds, Smithian.

13a–c: *Xiaoqiaoceras involutus* **n. gen., n. sp. PIMUZ 25948. Holotype.**
Loc. Jin4, Jinya, *Flemingites rursiradiatus* beds, Smithian.

14a–c: *Xiaoqiaoceras involutus* **n. gen., n. sp. PIMUZ 25949.**
Loc. Jin30, Jinya, *Flemingites rursiradiatus* beds, Smithian.

15a–c: *Xiaoqiaoceras involutus* **n. gen., n. sp. PIMUZ 25950.**
Loc. Jin30, Jinya, *Flemingites rursiradiatus* beds, Smithian.

16: **Suture line of** *Xiaoqiaoceras involutus* **n. gen., n. sp., PIMUZ 25951.**
Loc. Jin30, Jinya, *Flemingites rursiradiatus* beds, Smithian.
Scale bar = 5 mm; H = 9 mm.

17: *Parussuria compressa* **(Hyatt & Smith, 1905). PIMUZ 25952.**
Loc. Jin27, Jinya, *Owenites koeneni* beds, Smithian.

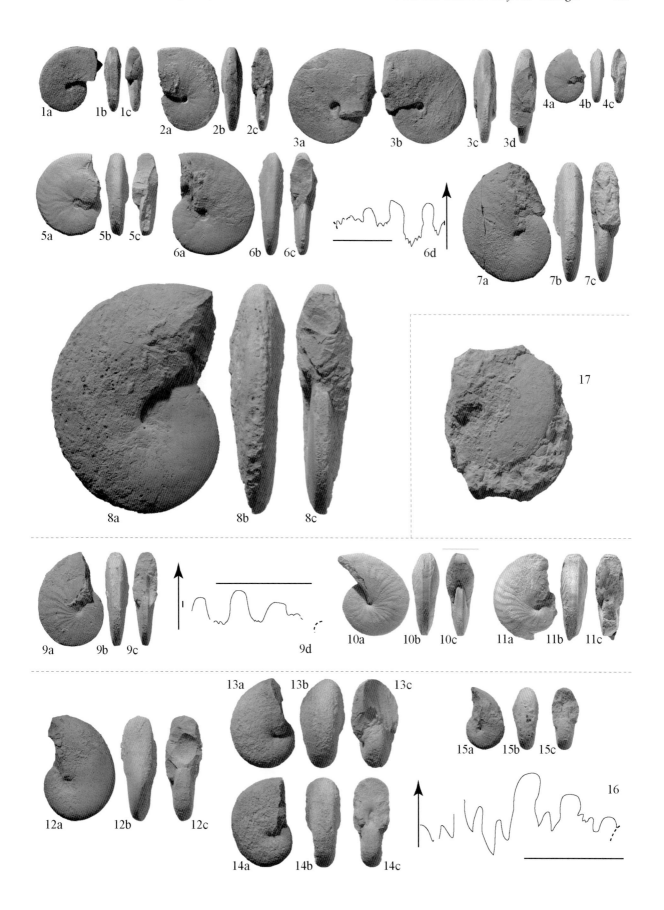

Plate 13
(All figures natural size)

1a–d: ***Wailiceras aemulus* n. gen., n. sp. PIMUZ 25953. Holotype.**
Loc. Jin61, Waili, *Kashmirites kapila* beds, Smithian.
a–c) Lateral, ventral and apertural views.
d) Suture line. Scale bar = 5 mm; H = 20 mm.

2a–c: ***Wailiceras aemulus* n. gen., n. sp. PIMUZ 25954.**
Loc. Jin64, Waili, *Kashmirites kapila* beds, Smithian.

3a–c: ***Wailiceras aemulus* n. gen., n. sp. PIMUZ 25955.**
Loc. Jin64, Waili, *Kashmirites kapila* beds, Smithian.

4a–b: ***Wailiceras aemulus* n. gen., n. sp. PIMUZ 25956.**
Loc. Jin64, Waili, *Kashmirites kapila* beds, Smithian.

5a–c: ***Wailiceras aemulus* n. gen., n. sp. PIMUZ 25957.**
Loc. Jin64, Waili, *Kashmirites kapila* beds, Smithian.

6a–c: ***Wailiceras aemulus* n. gen., n. sp. PIMUZ 25958.**
Loc. Jin64, Waili, *Kashmirites kapila* beds, Smithian.

7a–e: ***Wailiceras aemulus* n. gen., n. sp. PIMUZ 25959.**
Loc. Jin68, Waili, *Kashmirites kapila* beds, Smithian.
a–d) Lateral, ventral and apertural views.
e) Suture line. Scale bar = 1 mm; H = 2.5 mm. Very small specimen.

8a–c: ***Wailiceras aemulus* n. gen., n. sp. PIMUZ 25960.**
Loc. Jin65, Waili, *Kashmirites kapila* beds, Smithian.

9a–c: ***Wailiceras aemulus* n. gen., n. sp. PIMUZ 25961.**
Loc. Jin66, Waili, *Kashmirites kapila* beds, Smithian.

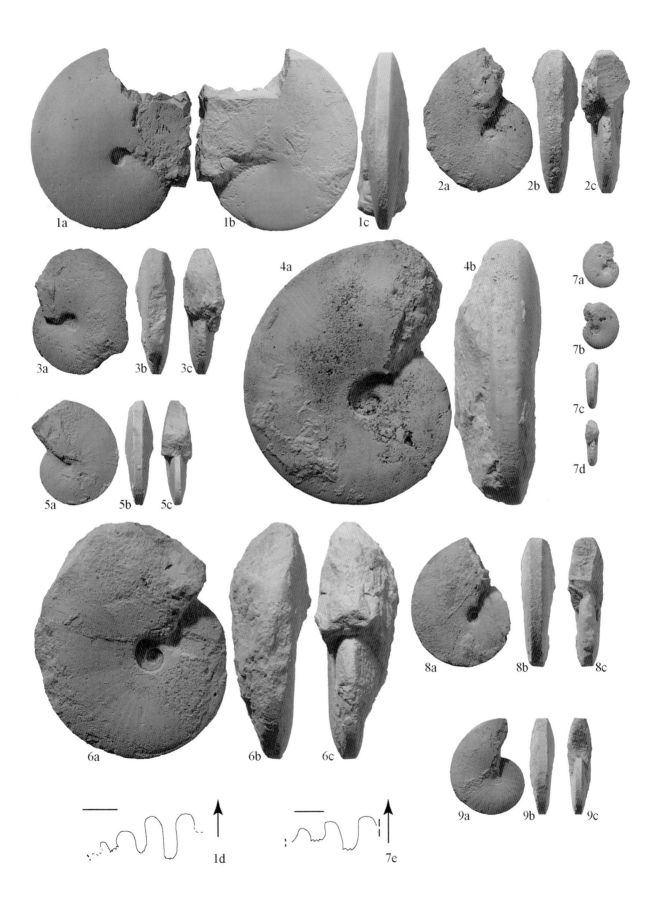

Plate 14
(All figures natural size)

1a–d: ***Leyeceras rothi*** **n. gen., n. sp. PIMUZ 25962.**
  Loc. Jin27, Jinya, *Owenites koeneni* beds, Smithian.

2a–e: ***Leyeceras rothi*** **n. gen., n. sp. PIMUZ 25963. Paratype.**
  Loc. Jin12, Jinya, *Owenites koeneni* beds, Smithian.
  a–d) Lateral, ventral and apertural views.
  e) Suture line. Scale bar = 5 mm; H = 18 mm.

3a–c: ***Leyeceras rothi*** **n. gen., n. sp. PIMUZ 25964. Holotype.**
  Loc. Jin43, Jinya, *Owenites koeneni* beds, Smithian.

4a–d: ***Urdyceras insolitus*** **n. gen., n. sp. PIMUZ 25965. Holotype.**
  Loc. Jin30, Jinya, *Flemingites rursiradiatus* beds, Smithian.
  a–d) Lateral, ventral and apertural views.
  e) Suture line. Scale bar = 5 mm; H = 10 mm.

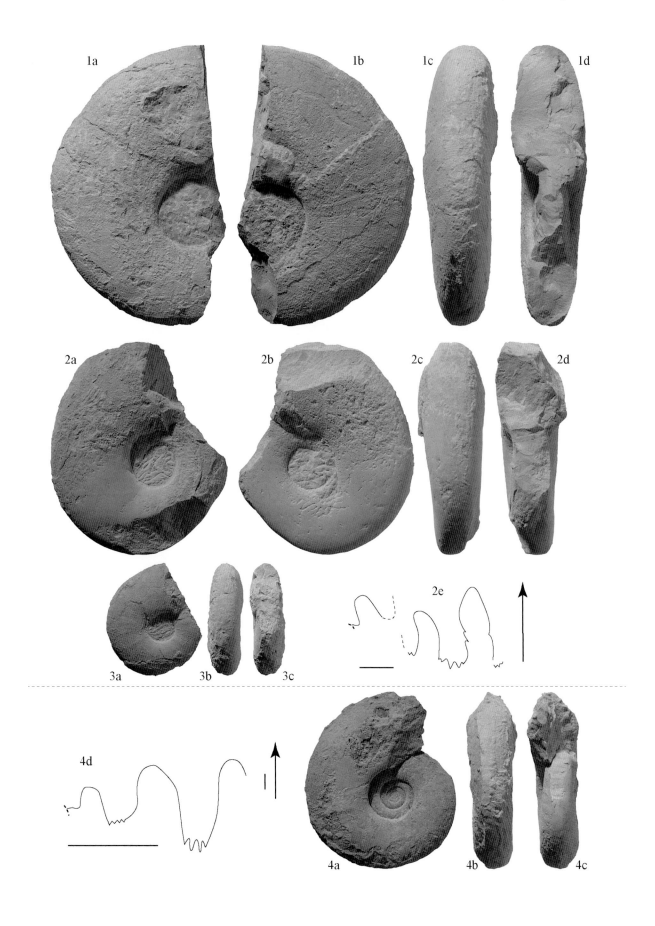

Plate 15
(All figures natural size)

1a–d: '*Gyronites*' cf. *superior* **Waagen, 1895. PIMUZ 25966.**
Loc. Jin61, Waili, *Kashmirites kapila* beds, Smithian.
a–c) Lateral, ventral and apertural views.
d) Suture line. Scale bar = 5 mm; H = 21 mm.

2a–d: '*Gyronites*' cf. *superior* **Waagen, 1895. PIMUZ 25967.**
Loc. Jin61, Waili, *Kashmirites kapila* beds, Smithian.

3a–c: '*Gyronites*' cf. *superior* **Waagen, 1895. PIMUZ 25968.**
Loc. Jin66, Waili, *Kashmirites kapila* beds, Smithian.

4a–c: *Dieneroceras tientungense* **Chao, 1959. PIMUZ 25969.**
Loc. Jin13, Jinya, *Flemingites rursiradiatus* beds, Smithian.

5a–d: *Dieneroceras tientungense* **Chao, 1959. PIMUZ 25970.**
Loc. Jin13, Jinya, *Flemingites rursiradiatus* beds, Smithian.

6a–c: *Dieneroceras tientungense* **Chao, 1959. PIMUZ 25971.**
Loc. Jin13, Jinya, *Flemingites rursiradiatus* beds, Smithian.

7a–c: *Dieneroceras tientungense* **Chao, 1959. PIMUZ 25972.**
Loc. Jin30, Jinya, *Flemingites rursiradiatus* beds, Smithian.

8a–c: *Dieneroceras tientungense* **Chao, 1959. PIMUZ 25973.**
Loc. Jin29, Jinya, *Flemingites rursiradiatus* beds, Smithian.

9a–c: *Dieneroceras tientungense* **Chao, 1959. PIMUZ 25974.**
Loc. Jin30, Jinya, *Flemingites rursiradiatus* beds, Smithian.

10a–b: *Dieneroceras tientungense* **Chao, 1959. PIMUZ 25975.**
Loc. Jin30, Jinya, *Flemingites rursiradiatus* beds, Smithian.

11:    **Suture line of** *Dieneroceras tientungense* **Chao, 1959. PIMUZ 25976.**
Loc. Jin30, Jinya, *Flemingites rursiradiatus* beds, Smithian.
Scale bar = 5 mm; D = 25 mm.

12:    **Suture line of** *Dieneroceras tientungense* **Chao, 1959. PIMUZ 25977.**
Loc. Jin28, Jinya, *Flemingites rursiradiatus* beds, Smithian.
Scale bar = 5 mm; D = 40 mm.

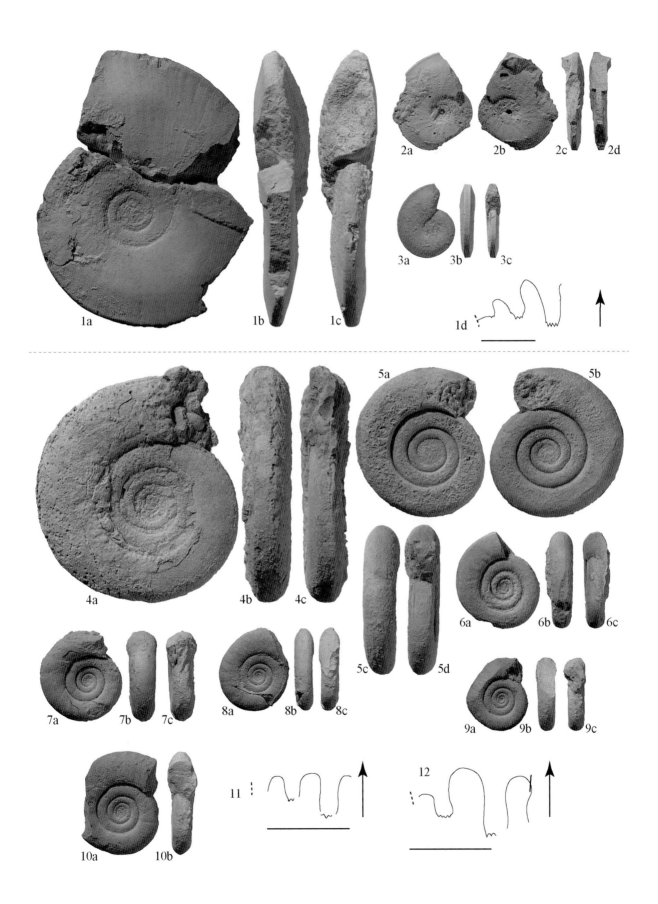

Plate 16
(All figures natural size)

1a–d: *Wyomingites aplanatus* **(White, 1879). PIMUZ 25978.**
Loc. Jin28, Jinya, *Flemingites rursiradiatus* beds, Smithian.
a–c) Lateral, ventral and apertural views.
d) Suture line. Scale bar = 5 mm; H = 8 mm.

2a–c: *Wyomingites aplanatus* **(White, 1879). PIMUZ 25979.**
Loc. Jin30, Jinya, *Flemingites rursiradiatus* beds, Smithian.

3a–c: *Wyomingites aplanatus* **(White, 1879). PIMUZ 25980.**
Loc. Jin30, Jinya, *Flemingites rursiradiatus* beds, Smithian.

4a–c: *Submeekoceras mushbachanum* **(White, 1879). PIMUZ 25981.**
Loc. Jin30, Jinya, *Flemingites rursiradiatus* beds, Smithian.

Plate 17
(All figures natural size unless otherwise indicated)

1a–c: ***Flemingites flemingianus* (de Koninck, 1863). PIMUZ 25982.**
Loc. Jin30, Jinya, *Flemingites rursiradiatus* beds, Smithian.

2a–c: ***Flemingites flemingianus* (de Koninck, 1863). PIMUZ 25983. Scale ×0.5.**
Loc. Jin15, Jinya, *Flemingites rursiradiatus* beds, Smithian.

3a–c: ***Flemingites flemingianus* (de Koninck, 1863). PIMUZ 25984.**
Loc. Jin30, Jinya, *Flemingites rursiradiatus* beds, Smithian.

4a–c: ***Flemingites flemingianus* (de Koninck, 1863). PIMUZ 25985.**
Loc. Jin30, Jinya, *Flemingites rursiradiatus* beds, Smithian.

5a–d: ***Flemingites flemingianus* (de Koninck, 1863). PIMUZ 25986.**
Loc. FW5, Waili, *Flemingites rursiradiatus* beds, Smithian.
a–c) Lateral, ventral and apertural views.
d) Suture line. Scale bar = 5 mm; H = 20 mm.

Plate 18
(All figures natural size unless otherwise indicated)

1a–c: *Flemingites rursiradiatus* **Chao, 1959. PIMUZ 25987. Scale ×0.5.**
Loc. Jin28, Jinya, *Flemingites rursiradiatus* beds, Smithian.

2a–c: *Flemingites rursiradiatus* **Chao, 1959. PIMUZ 25988. Scale ×2.**
Loc. Jin30, Jinya, *Flemingites rursiradiatus* beds, Smithian.

3a–c: *Flemingites rursiradiatus* **Chao, 1959. PIMUZ 25989.**
Loc. Jin30, Jinya, *Flemingites rursiradiatus* beds, Smithian.

4a–c: *Flemingites rursiradiatus* **Chao, 1959. PIMUZ 25990.**
Loc. Jin29, Jinya, *Flemingites rursiradiatus* beds, Smithian.

5a–c: *Flemingites rursiradiatus* **Chao, 1959. PIMUZ 25991. Scale ×0.75.**
Loc. Jin30, Jinya, *Flemingites rursiradiatus* beds, Smithian.

6a–c: *Flemingites rursiradiatus* **Chao, 1959. PIMUZ 25992.**
Loc. Jin29, Jinya, *Flemingites rursiradiatus* beds, Smithian.

7: **Suture line of** *Flemingites rursiradiatus* **Chao, 1959. PIMUZ 25993.**
Loc. Jin41, Jinya, *Flemingites rursiradiatus* beds, Smithian.
Scale bar = 5 mm; H = 21 mm.

Plate 19
(All figures natural size)

1a–c: ***Flemingites rursiradiatus* Chao, 1959. PIMUZ 25994. Polygonal variant.**
Loc. Jin30, Jinya, *Flemingites rursiradiatus* beds, Smithian.

2a–c: ***Flemingites rursiradiatus* Chao, 1959. PIMUZ 25995. Polygonal variant.**
Loc. Jin28, Jinya, *Flemingites rursiradiatus* beds, Smithian.

3a–d: ***Flemingites rursiradiatus* Chao, 1959. PIMUZ 25996. Polygonal variant.**
Loc. Jin30, Jinya, *Flemingites rursiradiatus* beds, Smithian.
a–c) Lateral, ventral and apertural views.
d) Suture line. Scale bar = 5 mm; H = 6 mm.

4a–c: ***Flemingites radiatus* Waagen, 1895. PIMUZ 25997.**
Loc. Jin28, Jinya, *Flemingites rursiradiatus* beds, Smithian.

5a–c: ***Flemingites radiatus* Waagen, 1895. PIMUZ 25998.**
Loc. Jin28, Jinya, *Flemingites rursiradiatus* beds, Smithian.

6a–c: ***Flemingites radiatus* Waagen, 1895. PIMUZ 25999.**
Loc. Jin28, Jinya, *Flemingites rursiradiatus* beds, Smithian.

7a–c: ***Rohillites* sp. indet. PIMUZ 26000.**
Loc. Jin67, Waili, *Kashmirites kapila* beds, Smithian.

Plate 20
(All figures natural size)

1a–c: *Rohillites sobolevi* **n. sp. PIMUZ 26473. Holotype.**
Loc. FSB1/2, Jinya, *Flemingites rursiradiatus* beds, Smithian.

2a–c: *Rohillites sobolevi* **n. sp. PIMUZ 26474. Paratype.**
Loc. FSB1/2, Jinya, *Flemingites rursiradiatus* beds, Smithian.

3a–c: *Rohillites bruehwileri* **n. sp. PIMUZ 26475.**
Loc. FSB1/2, Jinya, *Flemingites rursiradiatus* beds, Smithian.

4a–c: *Rohillites bruehwileri* **n. sp. PIMUZ 26476.**
Loc. FSB1/2, Jinya, *Flemingites rursiradiatus* beds, Smithian.

5a–c: *Rohillites bruehwileri* **n. sp. PIMUZ 26477.**
Loc. FSB1/2, Jinya, *Flemingites rursiradiatus* beds, Smithian.

6a–c: *Rohillites bruehwileri* **n. sp. PIMUZ 26478. Holotype.**
Loc. Jin4, Jinya, *Flemingites rursiradiatus* beds, Smithian.

7a–c: *Rohillites bruehwileri* **n. sp. PIMUZ 26479.**
Loc. FSB1/2, Jinya, *Flemingites rursiradiatus* beds, Smithian.

8a–c: *Rohillites bruehwileri* **n. sp. PIMUZ 26480.**
Loc. FSB1/2, Jinya, *Flemingites rursiradiatus* beds, Smithian.

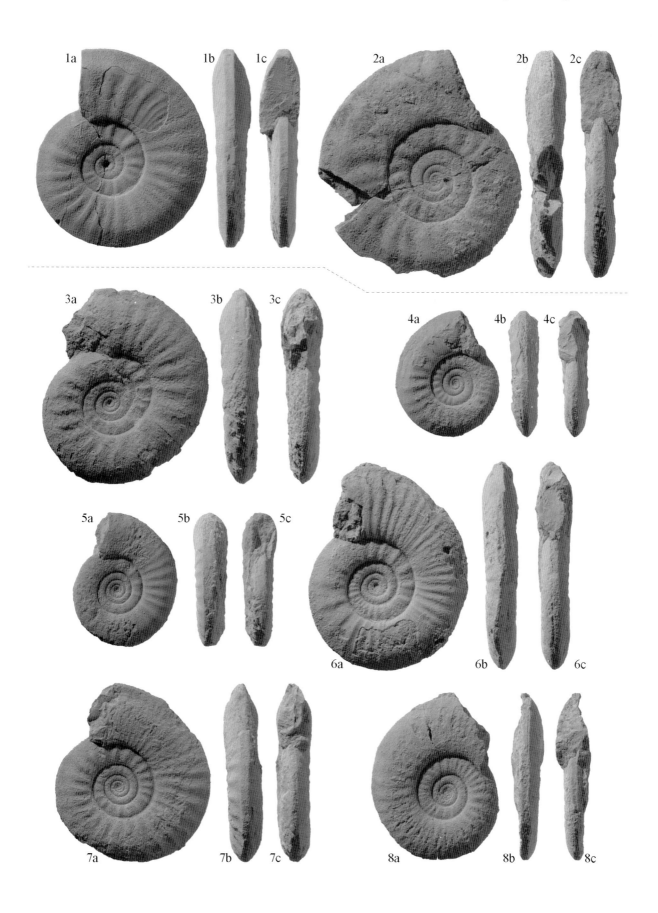

Plate 21
(All figures natural size)

1a–d: *Galfettites simplicitatis* **n. gen., n. sp. PIMUZ 26001. Paratype.**
Loc. Jin27, Jinya, *Owenites koeneni* beds, Smithian.
a–c) Lateral, ventral and apertural views.
d) Suture line. Scale bar = 5 mm; H = 23 mm.

2a–d: *Galfettites simplicitatis* **n. gen., n. sp. PIMUZ 26002. Holotype.**
Loc. Jin27, Jinya, *Owenites koeneni* beds, Smithian.
a–c) Lateral, ventral and apertural views.
d) Suture line. Scale bar = 5 mm; H = 30 mm.

1a   1b   1c   2d

1d

2a   2b   2c

Plate 22
(All figures natural size)

1a–b: *Pseudoflemingites goudemandi* **n. sp. PIMUZ 26003. Paratype.**
Loc. Jin99, Jinya, *Owenites koeneni* beds, Smithian.

2a–d: *Pseudoflemingites goudemandi* **n. sp. PIMUZ 26004. Holotype.**
Loc. Jin99, Jinya, *Owenites koeneni* beds, Smithian.
a–c) Lateral, ventral and apertural views.
d) Suture line. Scale bar = 5 mm; H = 12 mm.

3a–c: *Pseudoflemingites goudemandi* **n. sp. PIMUZ 26005. Paratype.**
Loc. Jin99, Jinya, *Owenites koeneni* beds, Smithian.

4a–c: *Pseudoflemingites goudemandi* **n. sp. PIMUZ 26006. Paratype.**
Loc. Jin99, Jinya, *Owenites koeneni* beds, Smithian.

5a–c: *Pseudoflemingites goudemandi* **n. sp. PIMUZ 26007. Paratype.**
Loc. Jin99, Jinya, *Owenites koeneni* beds, Smithian.

6a–d: *Juvenites procurvus* **n. sp. PIMUZ 26008.**
Loc. T5, Tsoteng, *Owenites koeneni* beds, Smithian.
a–c) Lateral, ventral and apertural views.
d) Suture line. Scale bar = 5 mm; H = 5 mm.

7a–c: *Juvenites procurvus* **n. sp. PIMUZ 26009.**
Loc. T5, Tsoteng, *Owenites koeneni* beds, Smithian.

8a–d: *Juvenites procurvus* **n. sp. PIMUZ 26010. Holotype.**
Loc. T11, Tsoteng, *Owenites koeneni* beds, Smithian.

9a–c: *Juvenites procurvus* **n. sp. PIMUZ 26011.**
Loc. Yu3, Yuping, *Owenites koeneni* beds, Smithian.

10a–c: *Juvenites procurvus* **n. sp. PIMUZ 26012.**
Loc. Yu3, Yuping, *Owenites koeneni* beds, Smithian.

11a–c: *Juvenites procurvus* **n. sp. PIMUZ 26013.**
Loc. Jin45, Jinya, *Owenites koeneni* beds, Smithian.

12a–c: *Juvenites procurvus* **n. sp. PIMUZ 26014.**
Loc. Jin45, Jinya, *Owenites koeneni* beds, Smithian.

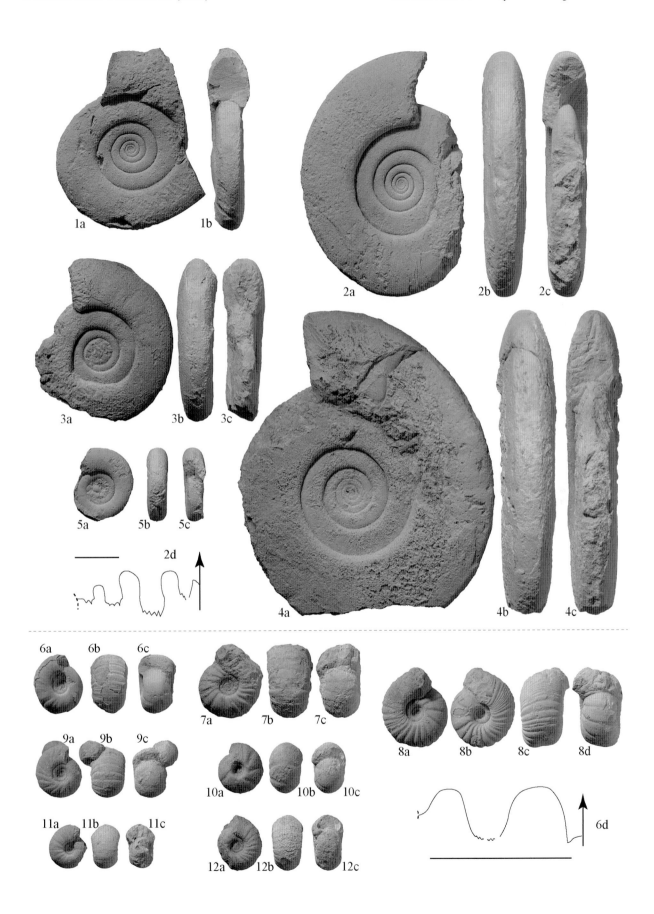

Plate 23
(All figures natural size unless otherwise indicated)

1a–b: **?*Anaxenaspis* sp. indet. PIMUZ 26015. Scale ×0.75.**
Loc. Jin45, Jinya, *Owenites koeneni* beds, Smithian.

2a–e: ***Guangxiceras inflata* n. gen., n. sp. PIMUZ 26016. Scale ×0.75. Holotype.**
Loc. Jin27, Jinya, *Owenites koeneni* beds, Smithian.
a–d) Lateral, ventral and apertural views.
e) Suture line. Scale bar = 5 mm; H = 25 mm.

FOSSILS AND STRATA 55 (2008)
              Smithian ammonoids from Guangxi
   135

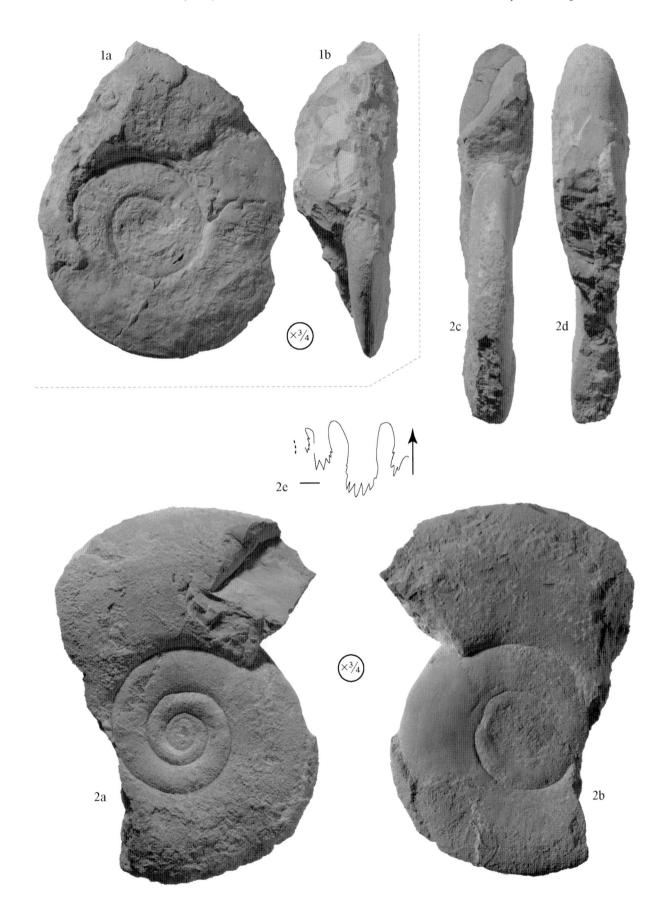

Plate 24
(All figures natural size unless otherwise indicated)

1a–d: *Larenites* cf. *reticulatus* **(Tozer, 1994). PIMUZ 26017.**
Loc. Jin66, Waili, *Kashmirites kapila* beds, Smithian.
1d: Scale ×3.

2a–c: *Larenites* cf. *reticulatus* **(Tozer, 1994). PIMUZ 26018.**
Loc. Jin66, Waili, *Kashmirites kapila* beds, Smithian.

3a–c: *Anaflemingites hochulii* **n. sp. PIMUZ 26019. Paratype.**
Loc. Jin45, Jinya, *Owenites koeneni* beds, Smithian.

4a–d: *Anaflemingites hochulii* **n. sp. PIMUZ 26020. Paratype.**
Loc. Jin45, Jinya, *Owenites koeneni* beds, Smithian.

5a–d: *Anaflemingites hochulii* **n. sp. PIMUZ 26021. Holotype.**
Loc. Jin45, Jinya, *Owenites koeneni* beds, Smithian.
a–c) Lateral, ventral and apertural views.
d) Suture line. Scale bar = 5 mm; H = 11 mm.

6a–c: *Anaflemingites hochulii* **n. sp. PIMUZ 26022. Paratype.**
Loc. Jin45, Jinya, *Owenites koeneni* beds, Smithian.

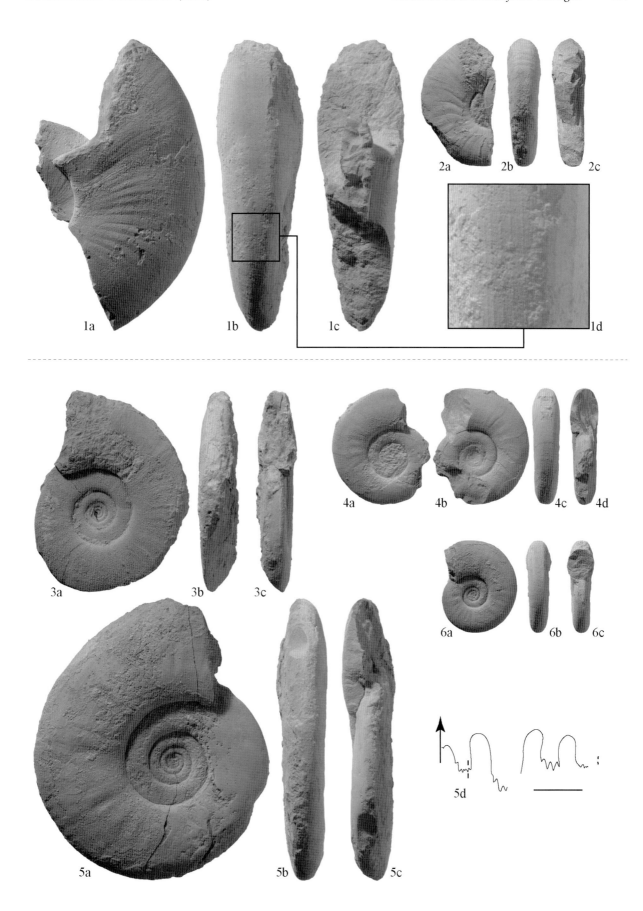

Plate 25
(All figures natural size)

1a–b: *Arctoceras strigatus* **n. sp. PIMUZ 26023. Holotype.**
Loc. Jin15, Jinya, *Flemingites rursiradiatus* beds, Smithian.

2a–d: *Arctoceras strigatus* **n. sp. PIMUZ 26024.**
Loc. FSB1/2, Jinya, *Flemingites rursiradiatus* beds, Smithian.
a–c) Lateral, ventral and apertural views.
d) Suture line. Scale bar = 5 mm; H = 12 mm.

3a–c: *Metussuria* **sp. indet. PIMUZ 26025. Scale ×0.75.**
Loc. Jin27, Jinya, *Owenites koeneni* beds, Smithian.

4a–b: *Metussuria* **sp. indet. PIMUZ 26026.**
Loc. Jin27, Jinya, *Owenites koeneni* beds, Smithian.

5a–b: *Metussuria* **sp. indet. PIMUZ 26027.**
Loc. Jin27, Jinya, *Owenites koeneni* beds, Smithian.

6a–b: *Metussuria* **sp. indet. PIMUZ 26028.**
Loc. Jin27, Jinya, *Owenites koeneni* beds, Smithian.

7: **Suture line of** *Metussuria* **sp. indet., PIMUZ 26029.**
Loc. Jin27, Jinya, *Owenites koeneni* beds, Smithian.
Scale bar = 5 mm; H = 37 mm.

8: **Suture line of** *Metussuria* **sp. indet., PIMUZ 26030.**
Loc. Jin27, Jinya, *Owenites koeneni* beds, Smithian.
Scale bar = 5 mm; H = 45 mm.

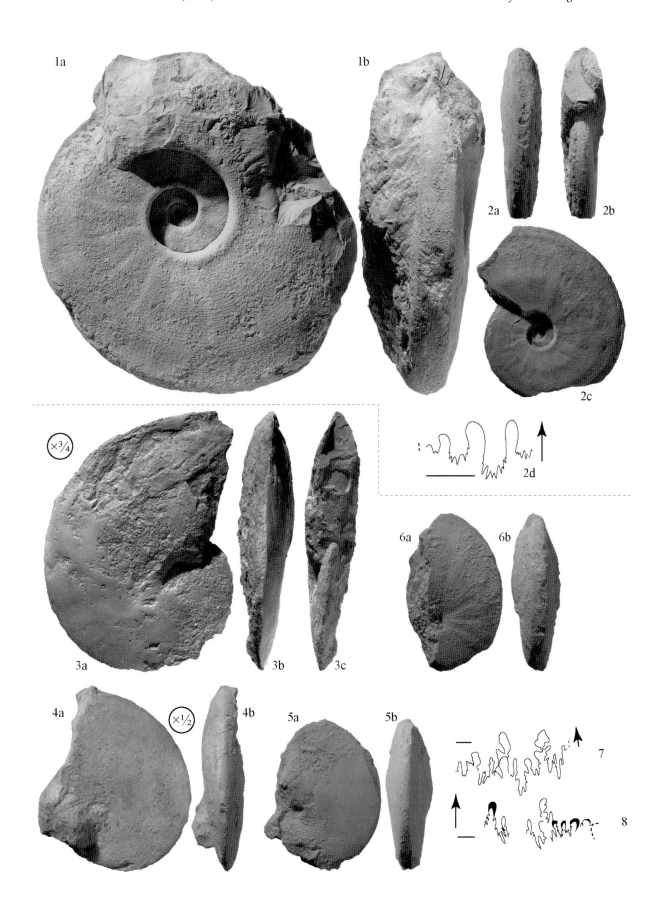

Plate 26
(All figures natural size unless otherwise indicated)

1a–d: *Submeekoceras mushbachanum* **(White, 1879). PIMUZ 26031.**
Loc. Jin28, Jinya, *Flemingites rursiradiatus* beds, Smithian.

2a–c: *Submeekoceras mushbachanum* **(White, 1879). PIMUZ 26032.**
Loc. Jin28, Jinya, *Flemingites rursiradiatus* beds, Smithian.

3a–c: *Submeekoceras mushbachanum* **(White, 1879). PIMUZ 26033.**
Loc. Jin28, Jinya, *Flemingites rursiradiatus* beds, Smithian.

4a–c: *Submeekoceras mushbachanum* **(White, 1879). PIMUZ 26034.**
Loc. Jin28, Jinya, *Flemingites rursiradiatus* beds, Smithian.

5a–d: *Submeekoceras mushbachanum* **(White, 1879). PIMUZ 26035.**
Loc. Jin28, Jinya, *Flemingites rursiradiatus* beds, Smithian.
a–c) Lateral, ventral and apertural views.
d) Suture line. Scale bar = 5 mm; H = 10 mm.

6a–c: *Submeekoceras mushbachanum* **(White, 1879). PIMUZ 26036.**
Loc. Jin28, Jinya, *Flemingites rursiradiatus* beds, Smithian.

7a–b: *Submeekoceras mushbachanum* **(White, 1879). PIMUZ 26037. Scale ×0.75.**
Loc. Jin13, Jinya, *Flemingites rursiradiatus* beds, Smithian.

8a–c: *Submeekoceras mushbachanum* **(White, 1879). PIMUZ 26038.**
Loc. Sha1, Shanggan, *Flemingites rursiradiatus* beds, Smithian.

9: **Suture line of** *Submeekoceras mushbachanum* **(White, 1879). PIMUZ 26039.**
Loc. Jin28, Jinya, *Flemingites rursiradiatus* beds, Smithian.
Scale bar = 5 mm; H = 25 mm.

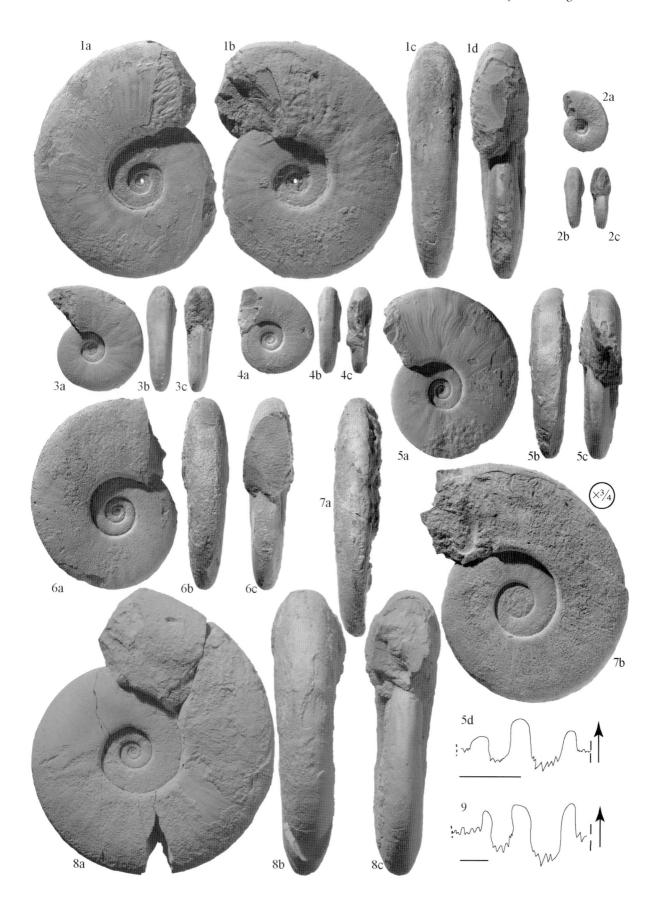

Plate 27
(All figures natural size unless otherwise indicated)

1a–d: ***Ussuria kwangsiana* Chao, 1959. PIMUZ 26040.**
Loc. Jin45, Jinya, *Owenites koeneni* beds, Smithian.
a–c) Lateral, ventral and apertural views. Scale ×0.75.
d) Suture line. Scale bar = 5 mm; H = 30 mm.

2a–c: ***Ussuria kwangsiana* Chao, 1959. PIMUZ 26041. Scale ×0.75.**
Loc. Jin45, Jinya, *Owenites koeneni* beds, Smithian.

3a–c: ***Ussuria kwangsiana* Chao, 1959. PIMUZ 26042. Scale ×0.75.**
Loc. Jin45, Jinya, *Owenites koeneni* beds, Smithian.

4a–c: ***Arctoceras* sp. indet. PIMUZ 26043. Scale ×0.5.**
Loc. FW5, Waili, *Flemingites rursiradiatus* beds, Smithian.

Plate 28
(All figures natural size unless otherwise indicated)

1a–b: *Anasibirites multiformis* **Welter, 1922. PIMUZ 26044. Scale ×0.75.**
Loc. Jin48, Jinya, *Anasibirites multiformis* beds, Smithian.

2a–c: *Anasibirites multiformis* **Welter, 1922. PIMUZ 26045.**
Loc. Jin48, Jinya, *Anasibirites multiformis* beds, Smithian.

3a–b: *Anasibirites multiformis* **Welter, 1922. PIMUZ 26046.**
Loc. Jin101, Jinya, *Anasibirites multiformis* beds, Smithian.

4a–c: *Anasibirites multiformis* **Welter, 1922. PIMUZ 26047.**
Loc. Jin101, Jinya, *Anasibirites multiformis* beds, Smithian.

5a–b: *Anasibirites multiformis* **Welter, 1922. PIMUZ 26048.**
Loc. FW6, Waili, *Anasibirites multiformis* beds, Smithian.

6: **Suture line of *Anasibirites multiformis* Welter, 1922. PIMUZ 26049.**
Loc. FW6, Waili, *Anasibirites multiformis* beds, Smithian.
Scale bar = 5 mm; H = 13 mm.

7a–c: *Anasibirites nevolini* **Burij & Zharnikova, 1968. PIMUZ 26050.**
Loc. Jin16, Jinya, *Anasibirites multiformis* beds, Smithian.

8a–d: *Anasibirites nevolini* **Burij & Zharnikova, 1968. PIMUZ 26051.**
Loc. FW6, Waili, *Anasibirites multiformis* beds, Smithian.

9a–c: *Anasibirites nevolini* **Burij & Zharnikova, 1968. PIMUZ 26052.**
Loc. FW6, Waili, *Anasibirites multiformis* beds, Smithian.

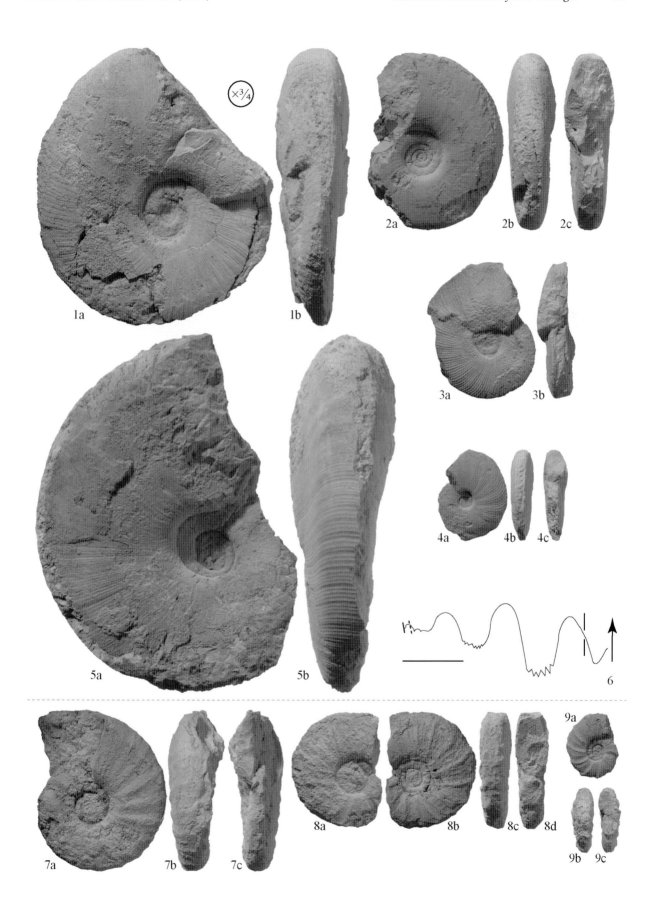

Plate 29
(All figures natural size)

1a–c: *Hemiprionites* **cf.** *butleri* **(Mathews, 1929). PIMUZ 26053.**
Loc. Jin48, Jinya, *Anasibirites multiformis* beds, Smithian.

2a–e: *Hemiprionites* **cf.** *butleri* **(Mathews, 1929). PIMUZ 26054.**
Loc. Jin48, Jinya, *Anasibirites multiformis* beds, Smithian.
a–d) Lateral, ventral and apertural views.
e) Suture line. Scale bar = 5 mm; H = 12 mm.

3a–c: *Hemiprionites* **cf.** *butleri* **(Mathews, 1929). PIMUZ 26055.**
Loc. Jin48, Jinya, *Anasibirites multiformis* beds, Smithian.

4a–c: *Hemiprionites* **cf.** *butleri* **(Mathews, 1929). PIMUZ 26056.**
Loc. Jin48, Jinya, *Anasibirites multiformis* beds, Smithian.

5a–c: *Hemiprionites* **cf.** *butleri* **(Mathews, 1929). PIMUZ 26057.**
Loc. Jin48, Jinya, *Anasibirites multiformis* beds, Smithian.

6a–c: *Hemiprionites* **cf.** *butleri* **(Mathews, 1929). PIMUZ 26058.**
Loc. Jin16, Jinya, *Anasibirites multiformis* beds, Smithian.

7:   *Hemiprionites* **cf.** *butleri* **(Mathews, 1929). PIMUZ 26059.**
Loc. Jin48, Jinya, *Anasibirites multiformis* beds, Smithian.

8a–d: *Subvishnuites stokesi* **(Kummel & Steele, 1962). PIMUZ 26060.**
Loc. Jin12, Jinya, *Owenites koeneni* beds, Smithian.
a–c) Lateral, ventral and apertural views.
d) Suture line. Scale bar = 5 mm; H = 13 mm.

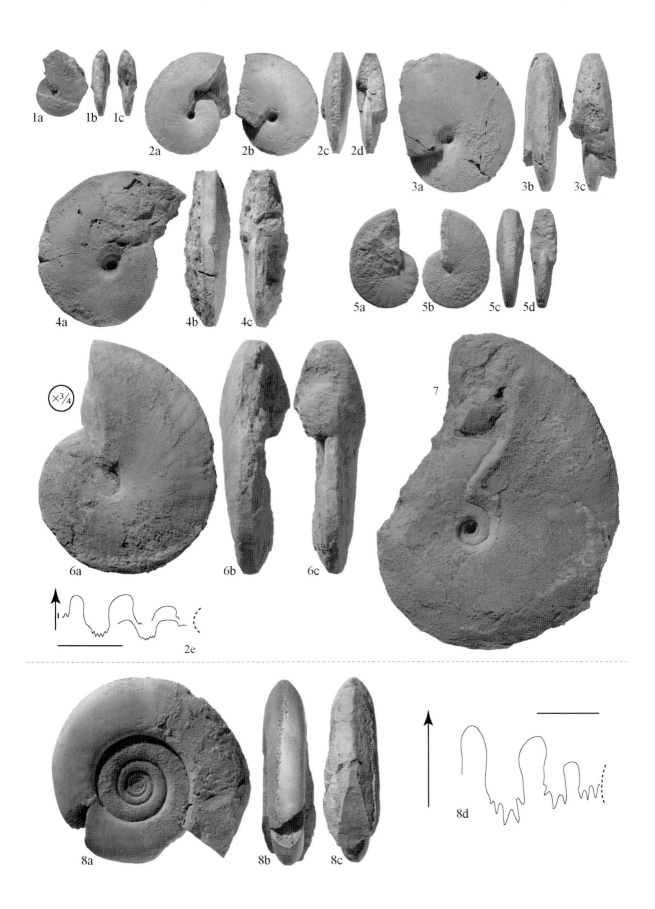

×3/4

Plate 30
(All figures natural size)

1a–c: ***Hemiprionites klugi* n. sp. PIMUZ 26061. Paratype.**
Loc. Jin16, Waili, *Anasibirites multiformis* beds, Smithian.

2a–c: ***Hemiprionites klugi* n. sp. PIMUZ 26062. Holotype.**
Loc. FW6, Waili, *Anasibirites multiformis* beds, Smithian.

3a–c: ***Hemiprionites klugi* n. sp. PIMUZ 26063.**
Loc. FW6, Jinya, *Anasibirites multiformis* beds, Smithian.

4:   **Suture line of *Hemiprionites klugi* n. sp. PIMUZ 26064.**
Loc. FW6, Waili, *Anasibirites multiformis* beds, Smithian.
Scale bar = 5 mm; H = 15 mm.

5a–c: ***Lanceolites compactus* Hyatt & Smith, 1905. PIMUZ 26065.**
Loc. Jin12, Jinya, *Owenites koeneni* beds, Smithian.

6a–d: ***Lanceolites bicarinatus* Smith, 1932. PIMUZ 26066.**
Loc. Yu7, Yuping, *Owenites koeneni* beds, Smithian.
a–c) Lateral, ventral and apertural views.
d) Suture line. Scale bar = 5 mm; H = 17 mm.

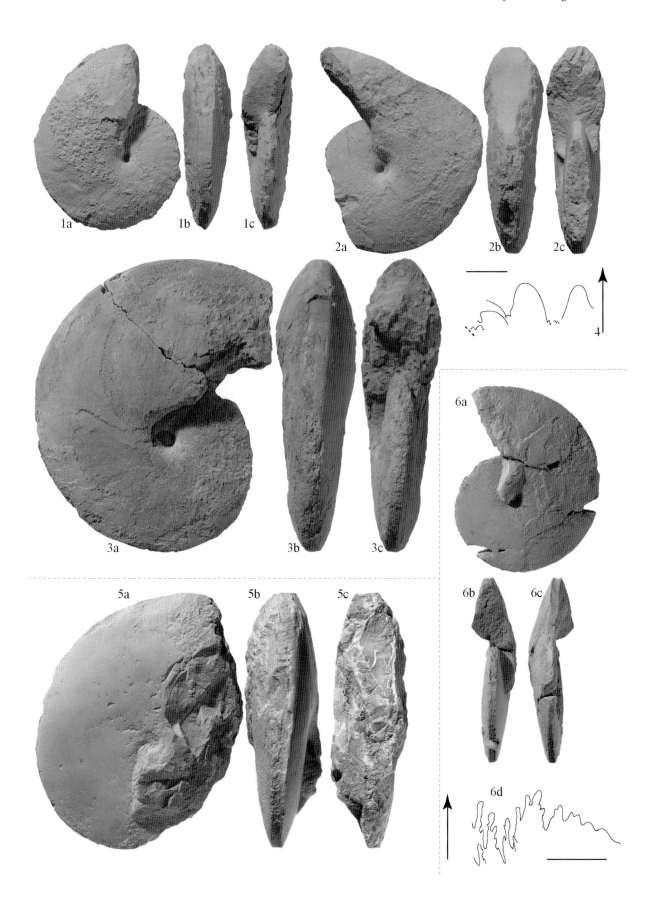

Plate 31
(All figures natural size)

1a–c: *Inyoites krystyni* **n. sp. PIMUZ 26067. Holotype.**
        Loc. Yu3, Yuping, *Owenites koeneni* beds, Smithian.

2a–d: *Inyoites krystyni* **n. sp. PIMUZ 26068. Paratype.**
        Loc. Yu3, Yuping, *Owenites koeneni* beds, Smithian.

3a–c: *Inyoites krystyni* **n. sp. PIMUZ 26069.**
        Loc. Jin12, Jinya, *Owenites koeneni* beds, Smithian.

4:      **Suture line of** *Inyoites krystyni* **n. sp., PIMUZ 26070.**
        Loc. Jin99, Jinya, *Owenites koeneni* beds, Smithian.
        Scale bar = 5 mm; H = 27 mm.

Plate 32
(All figures natural size unless otherwise indicated)

1a–c: *Inyoites krystyni* **n. sp. PIMUZ 26070. Scale ×0.75.**
Loc. Jin99, Jinya, *Owenites koeneni* beds, Smithian.

2a–c: *Inyoites krystyni* **n. sp. PIMUZ 26071. Scale ×0.75.**
Loc. Jin99, Jinya, *Owenites koeneni* beds, Smithian.

Plate 33
(All figures natural size unless otherwise indicated)

1a–c: *Paranannites* aff. *aspenensis* **Hyatt & Smith, 1905. PIMUZ 26072.**
Loc. Jin4, Jinya, *Flemingites rursiradiatus* beds, Smithian.

2a–c: *Paranannites* aff. *aspenensis* **Hyatt & Smith, 1905. PIMUZ 26073.**
Loc. FW5, Waili, *Flemingites rursiradiatus* beds, Smithian.

3a–c: *Paranannites* aff. *aspenensis* **Hyatt & Smith, 1905. PIMUZ 26074.**
Loc. Jin23, Jinya, *Flemingites rursiradiatus* beds, Smithian.

4a–c: *Paranannites* aff. *aspenensis* **Hyatt & Smith, 1905. PIMUZ 26075.**
Loc. Jin4, Jinya, *Flemingites rursiradiatus* beds, Smithian.

5a–c: *Paranannites* aff. *aspenensis* **Hyatt & Smith, 1905. PIMUZ 26076.**
Loc. FW5, Waili, *Flemingites rursiradiatus* beds, Smithian.

6a–c: *Paranannites* aff. *aspenensis* **Hyatt & Smith, 1905. PIMUZ 26077.**
Loc. Jin4, Jinya, *Flemingites rursiradiatus* beds, Smithian.

7a–d: *Paranannites* aff. *aspenensis* **Hyatt & Smith, 1905. PIMUZ 26078.**
Loc. Jin4, Jinya, *Flemingites rursiradiatus* beds, Smithian.
a–c) Lateral, ventral and apertural views.
d) Suture line. Scale bar = 5 mm; H = 3 mm.

8a–c: *Paranannites* aff. *aspenensis* **Hyatt & Smith, 1905. PIMUZ 26079.**
Loc. Jin30, Jinya, *Flemingites rursiradiatus* beds, Smithian.

9a–c: *Paranannites* aff. *aspenensis* **Hyatt & Smith, 1905. PIMUZ 26080.**
Loc. Jin28, Jinya, *Flemingites rursiradiatus* beds, Smithian.

10a–c: *Paranannites* aff. *aspenensis* **Hyatt & Smith, 1905. PIMUZ 26081.**
Loc. Jin4, Jinya, *Flemingites rursiradiatus* beds, Smithian.

11a–d: *Paranannites dubius* **n. sp. PIMUZ 26082. Scale ×2. Paratype.**
Loc. Jin4, Jinya, *Flemingites rursiradiatus* beds, Smithian.

12a–d: *Paranannites dubius* **n. sp. PIMUZ 26083. Scale ×2. Paratype.**
Loc. Jin4, Jinya, *Flemingites rursiradiatus* beds, Smithian.
a–c) Lateral, ventral and apertural views.
d) Suture line. Scale bar = 5 mm; H = 12 mm.

13a–c: *Paranannites dubius* **n. sp. PIMUZ 26084. Scale ×2. Holotype.**
Loc. Jin4, Jinya, *Flemingites rursiradiatus* beds, Smithian.

14a–c: *Paranannites dubius* **n. sp. PIMUZ 26085. Scale ×2. Paratype.**
Loc. Jin4, Jinya, *Flemingites rursiradiatus* beds, Smithian.

15a–d: Paranannitidae gen. indet. PIMUZ 26086.
Loc. Jin30, Jinya, *Flemingites rursiradiatus* beds, Smithian.
a–c) Lateral, ventral and apertural views.
d) Suture line. Scale bar = 5 mm; H = 3 mm.

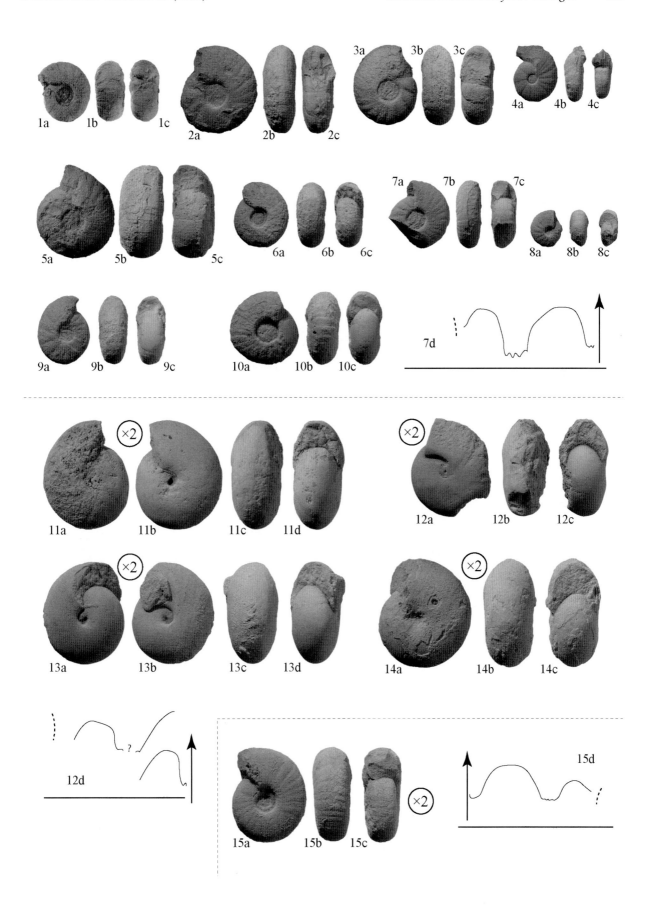

Plate 34
(All figures natural size)

1a–d: *'Paranannites' ovum* **n. sp. PIMUZ 26087. Paratype.**
Loc. Yu1, Yuping, *Owenites koeneni* beds, Smithian.

2a–d: *'Paranannites' ovum* **n. sp. PIMUZ 26088. Paratype.**
Loc. Yu1, Yuping, *Owenites koeneni* beds, Smithian.

3a–c: *'Paranannites' ovum* **n. sp. PIMUZ 26089. Holotype.**
Loc. Yu1, Yuping, *Owenites koeneni* beds, Smithian.

4a–b: *'Paranannites' ovum* **n. sp. PIMUZ 26090.**
Loc. T8, Tsoteng, *Owenites koeneni* beds, Smithian.

5a–c: *'Paranannites' ovum* **n. sp. PIMUZ 26091.**
Loc. T8, Tsoteng, *Owenites koeneni* beds, Smithian.

6:   **Suture line of *'Paranannites' ovum* n. sp., PIMUZ 26092.**
Loc. T8, Tsoteng, *Owenites koeneni* beds, Smithian.
Scale bar = 5 mm; H = 11 mm.

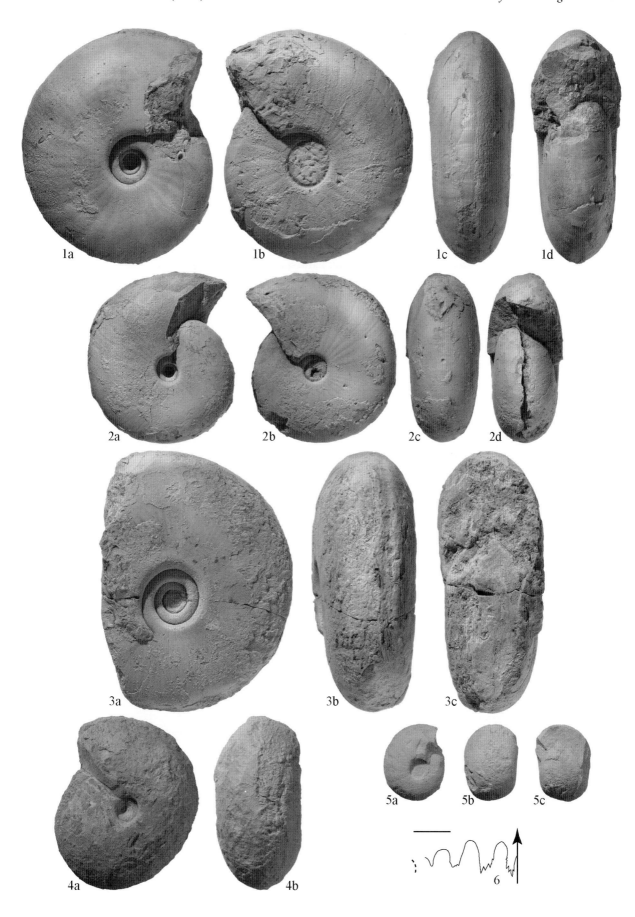

Plate 35
(All figures natural size)

1a–c: *Paranannites subangulosus* **n. sp. PIMUZ 26093.**
Loc. Jin29, Jinya, *Flemingites rursiradiatus* beds, Smithian.

2a–d: *Paranannites subangulosus* **n. sp. PIMUZ 26094. Holotype.**
Loc. Jin30, Jinya, *Flemingites rursiradiatus* beds, Smithian.
a–c) Lateral, ventral and apertural views.
d) Suture line. Scale bar = 5 mm; H = 13 mm.

3a–c: *Paranannites subangulosus* **n. sp. PIMUZ 26095. Paratype.**
Loc. Jin30, Jinya, *Flemingites rursiradiatus* beds, Smithian.

4a–c: *Paranannites subangulosus* **n. sp. PIMUZ 26096.**
Loc. Jin29, Jinya, *Flemingites rursiradiatus* beds, Smithian.

5a–c: *Paranannites subangulosus* **n. sp. PIMUZ 26097.**
Loc. Jin4, Jinya, *Flemingites rursiradiatus* beds, Smithian.

6a–c: *Paranannites subangulosus* **n. sp. PIMUZ 26098. Paratype.**
Loc. Jin30, Jinya, *Flemingites rursiradiatus* beds, Smithian.

7a–c: *Paranannites subangulosus* **n. sp. PIMUZ 26099. Paratype.**
Loc. Jin30, Jinya, *Flemingites rursiradiatus* beds, Smithian.

8a–c: *Paranannites subangulosus* **n. sp. PIMUZ 26100. Paratype.**
Loc. Jin30, Jinya, *Flemingites rursiradiatus* beds, Smithian.

9:   **Suture line of *Paranannites subangulosus* n. sp., PIMUZ 26101.**
Loc. Jin23, Jinya, *Flemingites rursiradiatus* beds, Smithian.
Scale bar = 5 mm; H = 4 mm.

10a–c: *Paranannites spathi* **(Frebold, 1930). PIMUZ 26102.**
Loc. Jin45, Jinya, *Owenites koeneni* beds, Smithian.

11a–c: *Paranannites spathi* **(Frebold, 1930). PIMUZ 26103.**
Loc. Jin45, Jinya, *Owenites koeneni* beds, Smithian.

12a–c: *Paranannites spathi* **(Frebold, 1930). PIMUZ 26104.**
Loc. Jin45, Jinya, *Owenites koeneni* beds, Smithian.

13a–c: *Paranannites spathi* **(Frebold, 1930). PIMUZ 26105.**
Loc. Yu1, Yuping, *Owenites koeneni* beds, Smithian.

14a–c: *Paranannites spathi* **(Frebold, 1930). PIMUZ 26106.**
Loc. Yu1, Yuping, *Owenites koeneni* beds, Smithian.

15a–c: *Paranannites spathi* **(Frebold, 1930). PIMUZ 26107.**
Loc. Yu1, Yuping, *Owenites koeneni* beds, Smithian.

16a–c: *Paranannites spathi* **(Frebold, 1930). PIMUZ 26108.**
Loc. Jin45, Jinya, *Owenites koeneni* beds, Smithian.

17a–c: *Paranannites spathi* **(Frebold, 1930). PIMUZ 26109.**
Loc. Yu1, Yuping, *Owenites koeneni* beds, Smithian.

18a–c: *Paranannites spathi* **(Frebold, 1930). PIMUZ 26110.**
Loc. Yu1, Yuping, *Owenites koeneni* beds, Smithian.

19:   **Suture line of *Paranannites spathi* (Frebold, 1930). PIMUZ 26111.**
Loc. Jin45, Jinya, *Owenites koeneni* beds, Smithian.
Scale bar = 5 mm; H = 5 mm.

20a–c: *Owenites simplex* **Welter, 1922. PIMUZ 26112.**
Loc. Jin45, Jinya, *Owenites koeneni* beds, Smithian.

21a–c: *Owenites simplex* **Welter, 1922. PIMUZ 26113.**
Loc. Jin45, Jinya, *Owenites koeneni* beds, Smithian.

22a–d: *Owenites simplex* **Welter, 1922. PIMUZ 26114.**
Loc. Jin45, Jinya, *Owenites koeneni* beds, Smithian.

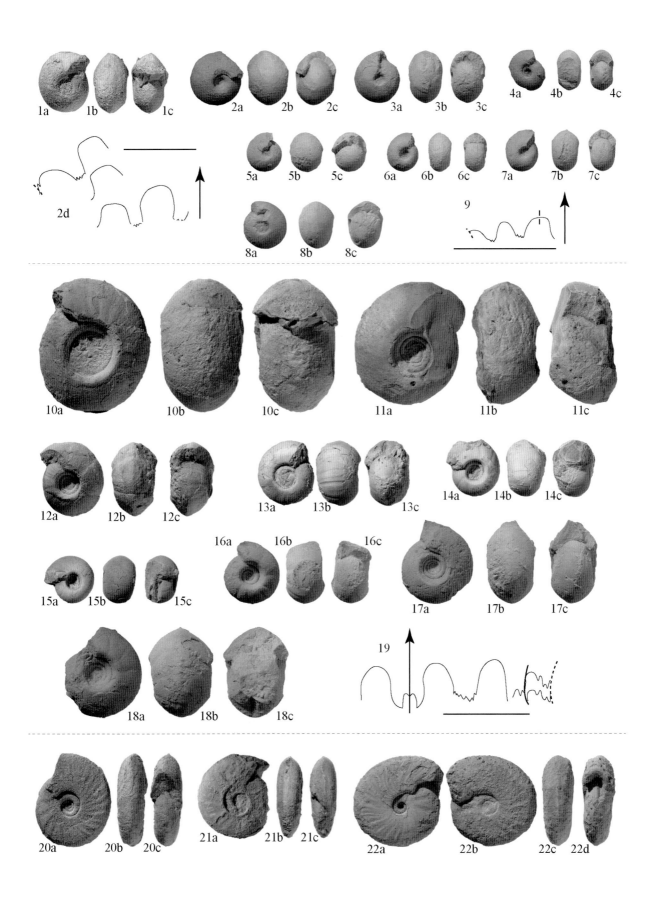

Plate 36
(All figures natural size)

1a–d: *Owenites koeneni* **Hyatt & Smith, 1905. PIMUZ 26115.**
Loc. T5, Tsoteng, *Owenites koeneni* beds, Smithian.

2a–c: *Owenites koeneni* **Hyatt & Smith, 1905. PIMUZ 26116.**
Loc. Jin27, Jinya, *Owenites koeneni* beds, Smithian.

3a–d: *Owenites koeneni* **Hyatt & Smith, 1905. PIMUZ 26117.**
Loc. T5, Tsoteng, *Owenites koeneni* beds, Smithian.

4a–d: *Owenites koeneni* **Hyatt & Smith, 1905. PIMUZ 26118.**
Loc. T5, Tsoteng, *Owenites koeneni* beds, Smithian.

5a–c: *Owenites koeneni* **Hyatt & Smith, 1905. PIMUZ 26119.**
Loc. Jin27, Jinya, *Owenites koeneni* beds, Smithian.

6a–d: *Owenites koeneni* **Hyatt & Smith, 1905. PIMUZ 26120.**
Loc. Jin99, Jinya, *Owenites koeneni* beds, Smithian.
a–c) Lateral, ventral and apertural views.
d) Suture line. Scale bar = 5 mm; D = 60 mm. Slightly smoothed.

7: **Suture line of** *Owenites koeneni* **Hyatt & Smith, 1905. PIMUZ 26121.**
Loc. T5, Tsoteng, *Owenites koeneni* beds, Smithian.
Scale bar = 5 mm; H = 12 mm.

8: **Suture line of** *Owenites koeneni* **Hyatt & Smith, 1905. PIMUZ 26122.**
Loc. Jin44, Jinya, *Owenites koeneni* beds, Smithian.
Scale bar = 5 mm; H = 25 mm. Slightly smoothed.

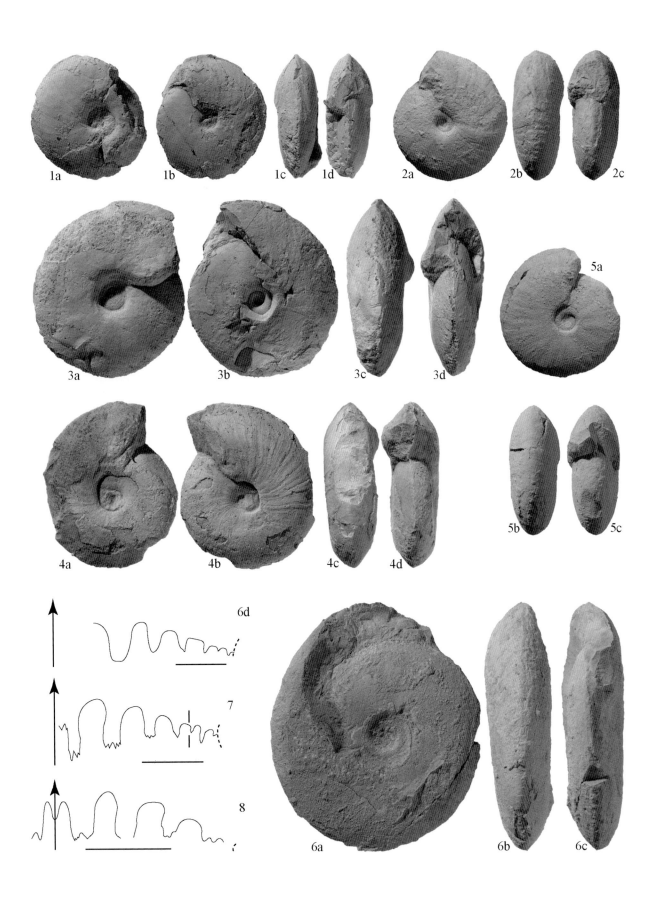

Plate 37
(All figures natural size unless otherwise indicated)

1a–d: ***Pseudosageceras multilobatum* Noetling, 1905. PIMUZ 26123.**
Loc. Jin30, Jinya, *Flemingites rursiradiatus* beds, Smithian.
a–c) Lateral, ventral and apertural views.
d) Suture line. Scale bar = 5 mm; H = 25 mm.

2a–c: ***Pseudosageceras multilobatum* Noetling, 1905. PIMUZ 26124.**
Loc. Jin30, Jinya, *Flemingites rursiradiatus* beds, Smithian.

3a–c: ***Pseudosageceras multilobatum* Noetling, 1905. PIMUZ 26125.**
Loc. Jin4, Jinya, *Flemingites rursiradiatus* beds, Smithian.

4a–c: ***Pseudosageceras multilobatum* Noetling, 1905. PIMUZ 26126.**
Loc. Jin30, Jinya, *Flemingites rursiradiatus* beds, Smithian.

5a–c: ***Pseudosageceras multilobatum* Noetling, 1905. PIMUZ 26127. Scale ×0.75.**
Loc. Jin4, Jinya, *Flemingites rursiradiatus* beds, Smithian.

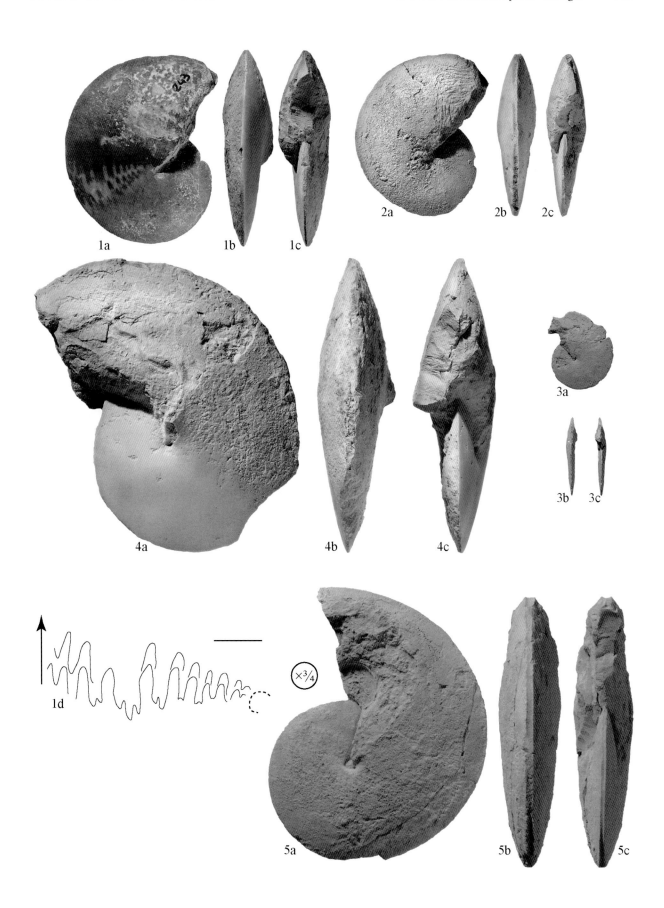

1a   1b   1c   2a   2b   2c

3a

3b   3c

4a   4b   4c

1d   ×3/4

5a   5b   5c

Plate 38
(All figures natural size unless otherwise indicated)

1a–c: *Clypites* **sp. indet. PIMUZ 26128.**
       Loc. Jin62, Waili, *Clypites* sp. indet. beds, Smithian.

2a–d: *Clypites* **sp. indet. PIMUZ 26129.**
       Loc. Jin62, Waili, *Clypites* sp. indet. beds, Smithian.
       a–c) Lateral, ventral and apertural views.
       d) Suture line. Scale bar = 5 mm; H = 11 mm. Umbilical part.

3a–c: *Clypites* **sp. indet. PIMUZ 26130.**
       Loc. Jin62, Waili, *Clypites* sp. indet. beds, Smithian.

4a–c: *Clypites* **sp. indet. PIMUZ 26131.**
       Loc. Jin62, Waili, *Clypites* sp. indet. beds, Smithian.

5a–d: *Proharpoceras carinatitabulatum* **Chao, 1950. PIMUZ 26132.**
       Loc. Jin45, Jinya, *Owenites koeneni* beds, Smithian.

6a–d: *Proharpoceras carinatitabulatum* **Chao, 1950. PIMUZ 26133.**
       Loc. Jin45, Jinya, *Owenites koeneni* beds, Smithian.

7:     *Proharpoceras carinatitabulatum* **Chao, 1950. PIMUZ 26134.**
       Loc. Jin45, Jinya, *Owenites koeneni* beds, Smithian.

8a–c: *Proharpoceras carinatitabulatum* **Chao, 1950. PIMUZ 26135. Scale ×2.**
       Loc. Jin45, Jinya, *Owenites koeneni* beds, Smithian.

9a–d: *Proharpoceras carinatitabulatum* **Chao, 1950. PIMUZ 26136.**
       Loc. Yu1, Yuping, *Owenites koeneni* beds, Smithian.
       a–c) Lateral, ventral and apertural views.
       d) Suture line. Scale bar = 5 mm; H = 6 mm.

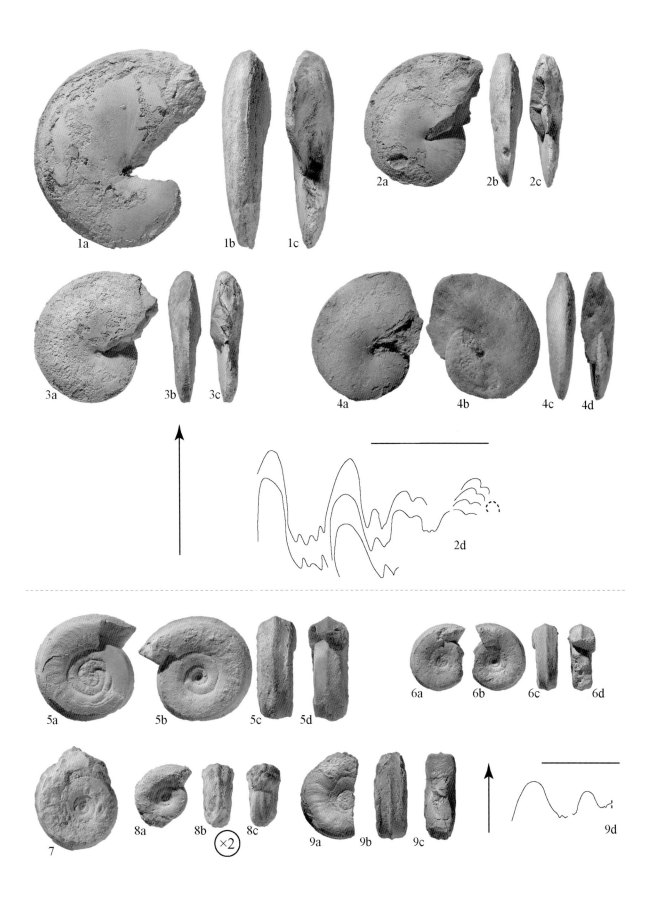

Plate 39
(All figures natural size unless otherwise indicated)

1a–b: *Hedenstroemia augusta* **n. sp. PIMUZ 26137. Paratype.**
         Loc. NW13, Waili, *Anasibirites multiformis* beds, Smithian.

2a–c: *Hedenstroemia augusta* **n. sp. PIMUZ 26138. Holotype.**
         Loc. NW13, Waili, *Anasibirites multiformis* beds, Smithian.

3a–c: *Hedenstroemia augusta* **n. sp. PIMUZ 26139. Paratype.**
         Loc. NW13, Waili, *Anasibirites multiformis* beds, Smithian.

4a–c: *Hedenstroemia augusta* **n. sp. PIMUZ 26140. Paratype. Scale ×2.**
         Loc. NW13, Waili, *Anasibirites multiformis* beds, Smithian.

5a–d: *Hedenstroemia augusta* **n. sp. PIMUZ 26141. Paratype. Scale ×2.**
         Loc. NW13, Waili, *Anasibirites multiformis* beds, Smithian.

6a–d: *Hedenstroemia augusta* **n. sp. PIMUZ 26142. Paratype. Scale ×2.**
         Loc. NW13, Waili, *Anasibirites multiformis* beds, Smithian.

7a–d: *Hedenstroemia augusta* **n. sp. PIMUZ 26143. Paratype.**
         Loc. NW13, Waili, *Anasibirites multiformis* beds, Smithian.

8a–c: *Hedenstroemia augusta* **n. sp. PIMUZ 26144. Paratype. Scale ×2.**
         Loc. NW13, Waili, *Anasibirites multiformis* beds, Smithian.

9a–d: *Hedenstroemia augusta* **n. sp. PIMUZ 26145. Paratype. Scale ×2.**
         Loc. NW13, Waili, *Anasibirites multiformis* beds, Smithian.

10a–c: *Hedenstroemia augusta* **n. sp. PIMUZ 26146.**
         Loc. Jin33, Jinya, *Anasibirites multiformis* beds, Smithian.

11:     **Suture line of** *Hedenstroemia augusta* **n. sp., PIMUZ 26147.**
         Loc. NW13, Waili, *Anasibirites multiformis* beds, Smithian.
         Scale bar = 5 mm; H = 20 mm.

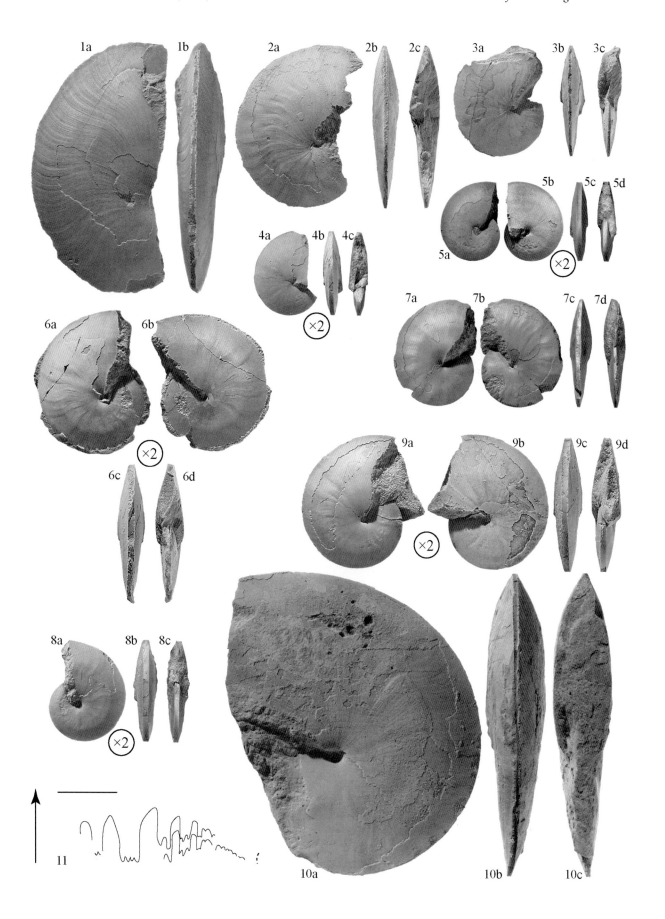

Plate 40
(All figures natural size unless otherwise indicated)

1a–e: ***Cordillerites antrum* n. sp. PIMUZ 26148. Holotype.**
Loc. Jin61, Waili, *Kashmirites kapila* beds, Smithian.
a–d) Lateral, ventral and apertural views.
e) Suture line. Scale bar = 5 mm; H = 13 mm.

2a–c: ***Cordillerites antrum* n. sp. PIMUZ 26149. Paratype.**
Loc. Jin61, Waili, *Kashmirites kapila* beds, Smithian.

3a–d: ***Cordillerites antrum* n. sp. PIMUZ 26150.**
Loc. Jin64, Waili, *Kashmirites kapila* beds, Smithian.

4a–c: ***Cordillerites antrum* n. sp. PIMUZ 26151.**
Loc. Jin64, Waili, *Kashmirites kapila* beds, Smithian.

5a–b: ***Cordillerites antrum* n. sp. PIMUZ 26152. Paratype.**
Loc. Jin61, Waili, *Kashmirites kapila* beds, Smithian.

6a–b: ***Cordillerites antrum* n. sp. PIMUZ 26153. Paratype.**
Loc. Jin61, Waili, *Kashmirites kapila* beds, Smithian.

7a–d: ***Cordillerites antrum* n. sp. PIMUZ 26154.**
Loc. Jin64, Waili, *Kashmirites kapila* beds, Smithian.
a–c) Lateral, ventral and apertural views. Scale ×0.75.
d) Suture line. Scale bar = 5 mm; H = 30 mm.

8: ***Cordillerites antrum* n. sp. PIMUZ 26155. Paratype.**
Loc. Jin61, Waili, *Kashmirites kapila* beds, Smithian.

9a–d: ***Cordillerites antrum* n. sp. PIMUZ 26156.**
Loc. Jin66, Waili, *Kashmirites kapila* beds, Smithian.

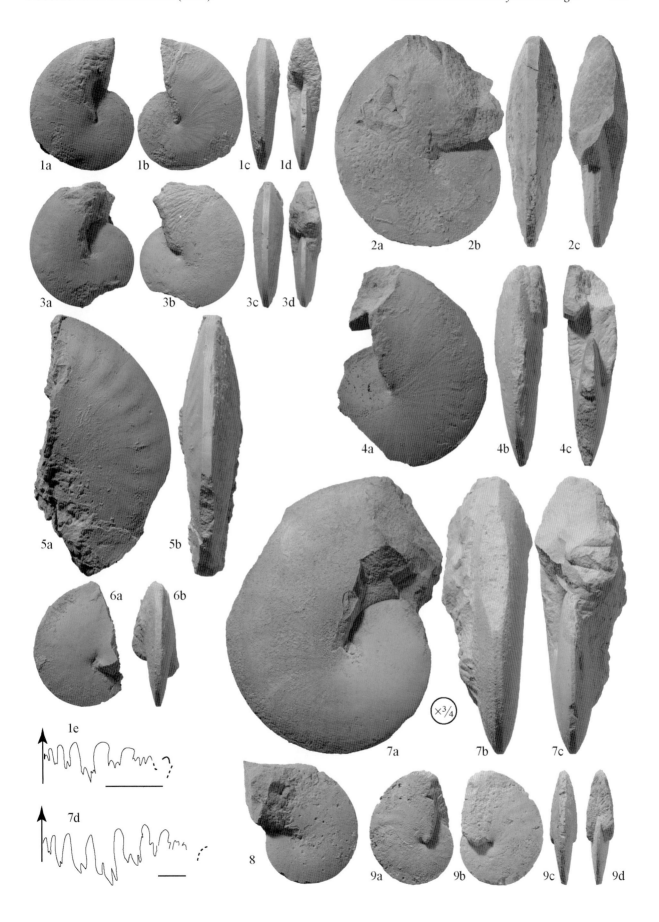

1a 1b 1c 1d

3a 3b 3c 3d

2a 2b 2c

5a 5b

4a 4b 4c

6a 6b

1e

7d

7a ×³⁄₄ 7b 7c

8

9a 9b 9c 9d

Plate 41
(All figures natural size)

1a–b: *Mesohedenstroemia kwangsiana* **Chao, 1959. PIMUZ 26157.**
Loc. Jin4, Jinya, *Flemingites rursiradiatus* beds, Smithian.

2a–c: *Mesohedenstroemia kwangsiana* **Chao, 1959. PIMUZ 26158.**
Loc. Jin4, Jinya, *Flemingites rursiradiatus* beds, Smithian.

3a–c: *Mesohedenstroemia kwangsiana* **Chao, 1959. PIMUZ 26159.**
Loc. Jin4, Jinya, *Flemingites rursiradiatus* beds, Smithian.

4a–c: *Mesohedenstroemia kwangsiana* **Chao, 1959. PIMUZ 26160.**
Loc. Jin4, Jinya, *Flemingites rursiradiatus* beds, Smithian.

5a–c: *Mesohedenstroemia kwangsiana* **Chao, 1959. PIMUZ 26161.**
Loc. Jin4, Jinya, *Flemingites rursiradiatus* beds, Smithian.

6a–c: *Mesohedenstroemia kwangsiana* **Chao, 1959. PIMUZ 26162.**
Loc. Jin29, Jinya, *Flemingites rursiradiatus* beds, Smithian.

7a–c: *Mesohedenstroemia kwangsiana* **Chao, 1959. PIMUZ 26163.**
Loc. Jin10, Jinya, *Flemingites rursiradiatus* beds, Smithian.

8: **Suture line of** *Mesohedenstroemia kwangsiana* **Chao, 1959. PIMUZ 26164.**
Loc. Jin28, Jinya, *Flemingites rursiradiatus* beds, Smithian.
Scale bar = 5 mm; H = 13 mm.

9a–c: *Mesohedenstroemia planata* **Chao, 1959. PIMUZ 26165.**
Loc. Jin45, Jinya, *Owenites koeneni* beds, Smithian.
a–b) lateral and ventral views.
c) Suture line. Scale bar = 5 mm; H = 15 mm.

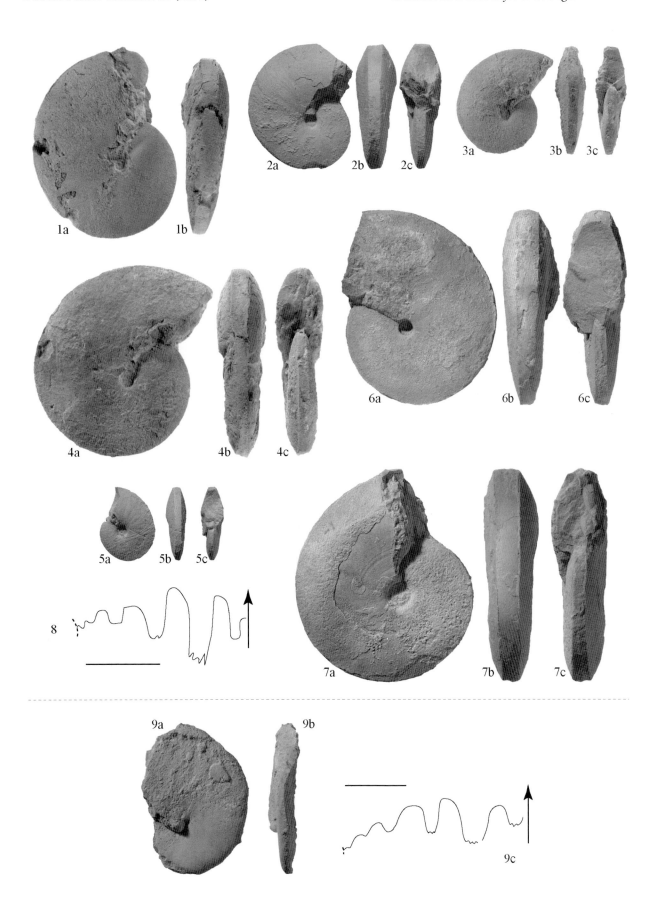

Plate 42
(All figures natural size)

1a–b: *Aspenites acutus* **Hyatt & Smith, 1905. PIMUZ 26166.**
Loc. Jin4, Jinya, *Flemingites rursiradiatus* beds, Smithian.

2a–c: *Aspenites acutus* **Hyatt & Smith, 1905. PIMUZ 26167.**
Loc. Jin4, Jinya, *Flemingites rursiradiatus* beds, Smithian.

3a–b: *Aspenites acutus* **Hyatt & Smith, 1905. PIMUZ 26168.**
Loc. Jin4, Jinya, *Flemingites rursiradiatus* beds, Smithian.

4a–c: *Aspenites acutus* **Hyatt & Smith, 1905. PIMUZ 26169.**
Loc. Jin28, Jinya, *Flemingites rursiradiatus* beds, Smithian.

5a–c: *Aspenites acutus* **Hyatt & Smith, 1905. PIMUZ 26170.**
Loc. Jin28, Jinya, *Flemingites rursiradiatus* beds, Smithian.

6a–c: *Aspenites acutus* **Hyatt & Smith, 1905. PIMUZ 26171.**
Loc. Jin28, Jinya, *Flemingites rursiradiatus* beds, Smithian.

7a–c: *Aspenites acutus* **Hyatt & Smith, 1905. PIMUZ 26172.**
Loc. Jin28, Jinya, *Flemingites rursiradiatus* beds, Smithian.

8: **Suture line of** *Aspenites acutus* **Hyatt & Smith, 1905. PIMUZ 26173.**
Loc. Jin27, Jinya, *Owenites koeneni* beds, Smithian.
Scale bar = 5 mm; H = 24 mm.

9: **Suture line of** *Aspenites acutus* **Hyatt & Smith, 1905. PIMUZ 26174.**
Loc. Jin27, Jinya, *Owenites koeneni* beds, Smithian.
Scale bar = 5 mm; H = 25 mm.

10a–b: **?***Aspenites* **sp. indet. PIMUZ 26175.**
Loc. Yu1, Yuping, *Owenites koeneni* beds, Smithian.

11a–b: **?***Aspenites* **sp. indet. PIMUZ 26176.**
Loc. NW1, Waili, *Owenites koeneni* beds, Smithian.

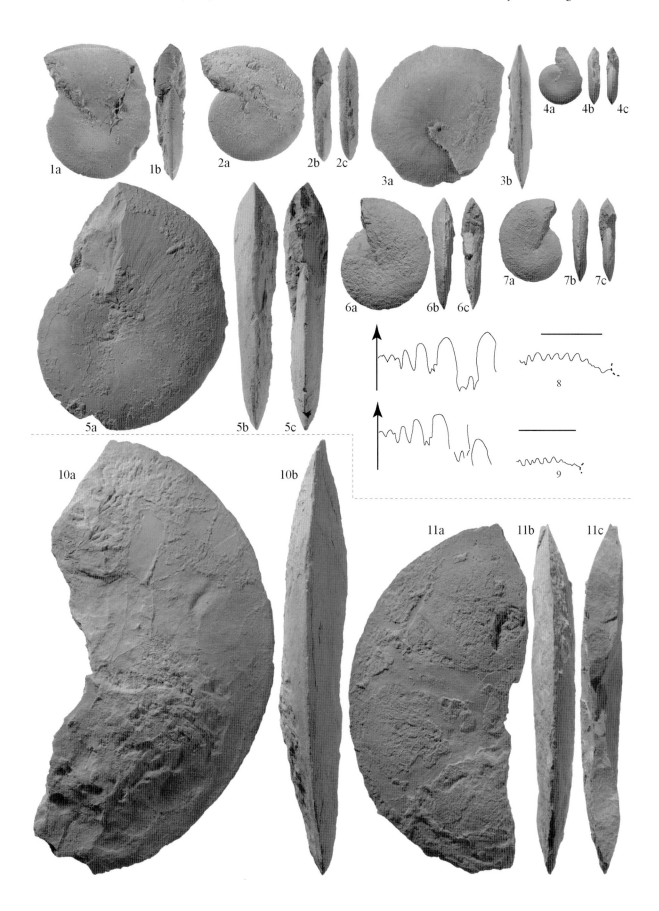

Plate 43
(All figures natural size)

1a–d: *Pseudaspenites layeriformis* **(Welter, 1922). PIMUZ 26177.**
Loc. Jin30, Jinya, *Flemingites rursiradiatus* beds, Smithian.

2a–c: *Pseudaspenites layeriformis* **(Welter, 1922). PIMUZ 26178.**
Loc. Jin30, Jinya, *Flemingites rursiradiatus* beds, Smithian.

3a–b: *Pseudaspenites layeriformis* **(Welter, 1922). PIMUZ 26179.**
Loc. Jin30, Jinya, *Flemingites rursiradiatus* beds, Smithian.

4a–d: *Pseudaspenites layeriformis* **(Welter, 1922). PIMUZ 26180.**
Loc. T50, Tsoteng, *Flemingites rursiradiatus* beds, Smithian.
a–c) Lateral, ventral and apertural views.
d) Suture line. Scale bar = 5 mm; H = 8 mm.

5a–c: *Pseudaspenites layeriformis* **(Welter, 1922). PIMUZ 26181.**
Loc. Jin29, Jinya, *Flemingites rursiradiatus* beds, Smithian.

6a–c: *Pseudaspenites layeriformis* **(Welter, 1922). PIMUZ 26182.**
Loc. Jin28, Jinya, *Flemingites rursiradiatus* beds, Smithian.

7a–c: *Pseudaspenites evolutus* **n. sp. PIMUZ 26183.**
Loc. Jin4, Jinya, *Flemingites rursiradiatus* beds, Smithian.

8a–c: *Pseudaspenites evolutus* **n. sp. PIMUZ 26184.**
Loc. Jin4, Jinya, *Flemingites rursiradiatus* beds, Smithian.

9a–c: *Pseudaspenites evolutus* **n. sp. PIMUZ 26185.**
Loc. Jin29, Jinya, *Flemingites rursiradiatus* beds, Smithian.

10a–d: *Pseudaspenites evolutus* **n. sp. PIMUZ 26186. Paratype.**
Loc. Jin30, Jinya, *Flemingites rursiradiatus* beds, Smithian.

11a–d: *Pseudaspenites evolutus* **n. sp. PIMUZ 26187. Holotype.**
Loc. Jin30, Jinya, *Flemingites rursiradiatus* beds, Smithian.

12a–e: *Pseudaspenites tenuis* **(Chao, 1959). PIMUZ 26188.**
Loc. Jin10, Jinya, *Flemingites rursiradiatus* beds, Smithian.
a–d) Lateral, ventral and apertural views.
e) Suture line. Scale bar = 5 mm; H = 15 mm.

13a–d: *Pseudaspenites tenuis* **(Chao, 1959). PIMUZ 26189.**
Loc. Jin30, Jinya, *Flemingites rursiradiatus* beds, Smithian.

14a–d: *Pseudaspenites tenuis* **(Chao, 1959). PIMUZ 26190.**
Loc. Jin30, Jinya, *Flemingites rursiradiatus* beds, Smithian.

15a–c: *Owenites carpenteri* **Smith, 1932. PIMUZ 26191.**
Loc. Jin47, Jinya, *Owenites koeneni* beds, Smithian.

16a–c: *Owenites carpenteri* **Smith, 1932. PIMUZ 26192.**
Loc. T12, Tsoteng, *Owenites koeneni* beds, Smithian.

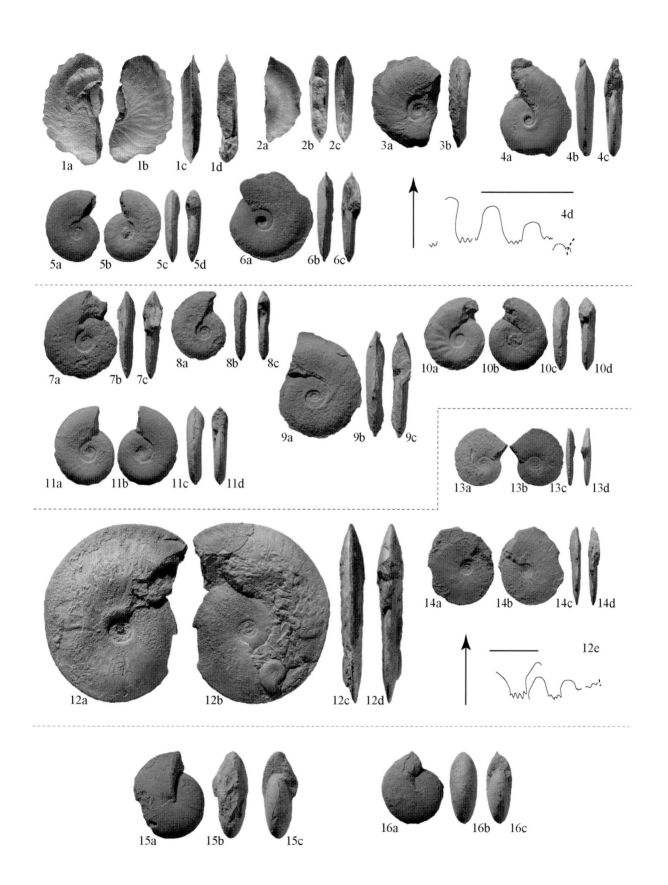

Plate 44
(All figures natural size unless otherwise indicated)

1a–d: ***Guodunites monneti* n. gen., n. sp. PIMUZ 26193. Holotype.**
Loc. Jin99, Jinya, *Owenites koeneni* beds, Smithian.
a–c) Lateral, ventral and apertural views. Scale ×0.5.
d) Suture line. Scale bar = 5 mm; H = 32 mm.

2a–d: ***Guodunites monneti* n. gen., n. sp. PIMUZ 26194.**
Loc. Jin12, Jinya, *Owenites koeneni* beds, Smithian.
a–c) Lateral, ventral and apertural views. Scale ×0.75.
d) Suture line. Scale bar = 5 mm; H = 35 mm.

3a–d: ***Procurvoceratites pygmaeus* n. gen., n. sp. PIMUZ 26195. Holotype. Scale ×2.**
Loc. Jin28, Jinya, *Flemingites rursiradiatus* beds, Smithian.

4a–b: ***Procurvoceratites pygmaeus* n. gen., n. sp. PIMUZ 26196. Scale ×2.**
Loc. Jin4, Jinya, *Flemingites rursiradiatus* beds, Smithian.

5a–b: ***Procurvoceratites pygmaeus* n. gen., n. sp. PIMUZ 26197. Scale ×2.**
Loc. Jin4, Jinya, *Flemingites rursiradiatus* beds, Smithian.

6a–c: ***Procurvoceratites ampliatus* n. gen., n. sp. PIMUZ 26198. Holotype. Scale ×2.**
Loc. Jin30, Jinya, *Flemingites rursiradiatus* beds, Smithian.

7a–c: ***Procurvoceratites tabulatus* n. gen., n. sp. PIMUZ 26199. Holotype. Scale ×2.**
Loc. Jin30, Jinya, *Flemingites rursiradiatus* beds, Smithian.

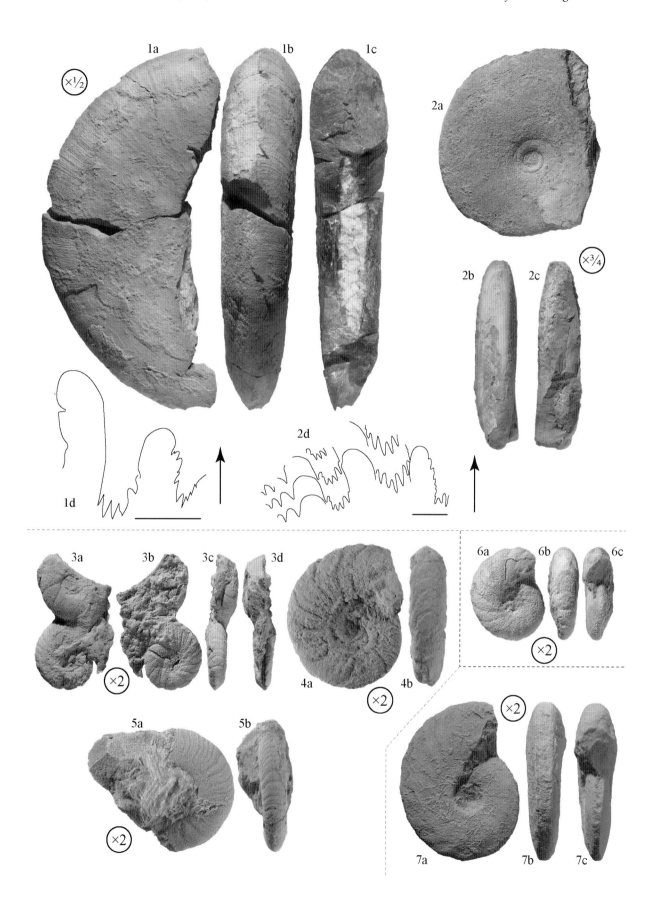

Plate 45
(All figures natural size)

1a–b: Gen. indet. A. PIMUZ 26200.
Loc. Jin12, Jinya, *Owenites koeneni* beds, Smithian.

2a–b: Gen. indet. C. PIMUZ 26201.
Loc. Yu22, Yuping, *Anasibirites multiformis* beds, Smithian.

3a–c: Gen. indet. D. PIMUZ 26202.
Loc. Yu22, Yuping, *Anasibirites multiformis* beds, Smithian.

4a–c: Gen. indet. B. PIMUZ 26203.
Loc. Yu22, Yuping, *Anasibirites multiformis* beds, Smithian.